The Hereford World Map

INTRODUCTION

The Hereford World Map

INTRODUCTION

P. D. A. Harvey

Hereford Cathedral
2010

Frontispiece: The Hereford world map, drawn on a single piece of parchment, 5 feet 2 inches high and 4 feet 4 inches wide (1.59 by 1.34 metres). Its colours, here restored, have changed over seven hundred years – the green seas are now dark brown, and the blue of the rivers has mostly flaked off

This edition is published by Hereford Cathedral with the kind and generous cooperation of The Folio Society

Printed in Hereford by
ABC PRINT

Contents

Preface

This is an expanded and updated version of *Mappa Mundi: The Hereford World Map*; this was first published in 1996 by Hereford Cathedral and the British Library and by the University of Toronto Press. It was reprinted with corrections in 1999, and a second edition was published in 2002 by Hereford Cathedral. Because the text has now been completely reset, it has been possible to take into account, more fully than in 2002, the proceedings of the 1999 conference (see p. 38), which contributed so much to our current understanding of the map, as well as other later research.

I would repeat here what I said in the preface to the first edition. It is a privilege that I much appreciate to have been invited by the Hereford Mappa Mundi Trustees to write this book and to have been given the opportunity of examining the map outside its case. In the course of the work many people have given me substantial help by way of advice and discussion, answering enquiries, correcting mistakes and providing generous hospitality, and I am especially grateful to Mr P. Barber, Dr A. D. Baynes-Cope, Mrs A. Butterworth, Canon J. Butterworth, Dr A. I. Doyle, Mr D. Harbour, Mr F. Herbert, Miss D. S. Hubbard, Dr M. Kupfer, Mr I. Milford, Mrs B. Morgan, Mr D. E. Morgan, Mrs A. Payne, Mme M. Pelletier, Dr N. Ramsay, Mr N. Skyrme, Dr S. D. Westrem and Miss J. Williams. I would add a special word of thanks to those with whom I have particularly worked, at Hereford Canon John Tiller, at the British Library Ms Kathleen Houghton and Mr David Way, and at The Folio Society Mr Joe Whitlock Blundell, Ms Karolina Maroszek and Dr Neil Titman.

P.D.A.H.
February 2010

I

The Map and Its Making

The Map Described

Hereford Cathedral's medieval map of the world is one of the most interesting artefacts to survive from thirteenth-century England, but at first sight the original map is far from prepossessing. It has aged over the centuries, and its pattern mostly of various shades of brown and black has a dull, uninteresting appearance. But when it was made, seven hundred years ago, it was alive with colour.

The parchment on which it is drawn would have been whiter then, and the dark brown seas that dominate the whole map were a bright green. It is easier to see that the rivers were once blue, probably bright blue, for though much of the colour has flaked off, leaving only the brown ground that held it to the parchment, it is in places almost intact and some traces remain throughout. The gold leaf has fared better: much of it remains, but it was austerely confined to some of the lettering – the names of the three continents and India on the map, the names of the four cardinal points in a circle around it and the letters MORS in roundels outside. The red no longer stands out as it must once have done, but it is still intact and bright; it is used for some decoration and lettering – including the names of peoples and provinces on the map and the wording round the edge of the parchment – and, as on some other medieval maps, not only for the Red Sea but for the Persian Gulf as well. The ink used for the outlines and most of the wording is mostly still crisp and black, unlike the ink of many contemporary manuscripts. Probably several other colours were used for the shading in the pictures on the map itself and in the borders: what now appear as grey, yellow and olive-brown can be distinguished.[1]

The Folio Society's reproduction of the map recaptures some of the lost colouring, but only so far as this can be done with certainty of its accuracy and without impinging on the map's other contents. The blue of the rivers has been restored, and we now see the seas on the map and the leaves in the top border in the bright green of the original paint. The red has been strengthened, and so has the gold. The brown background of parchment

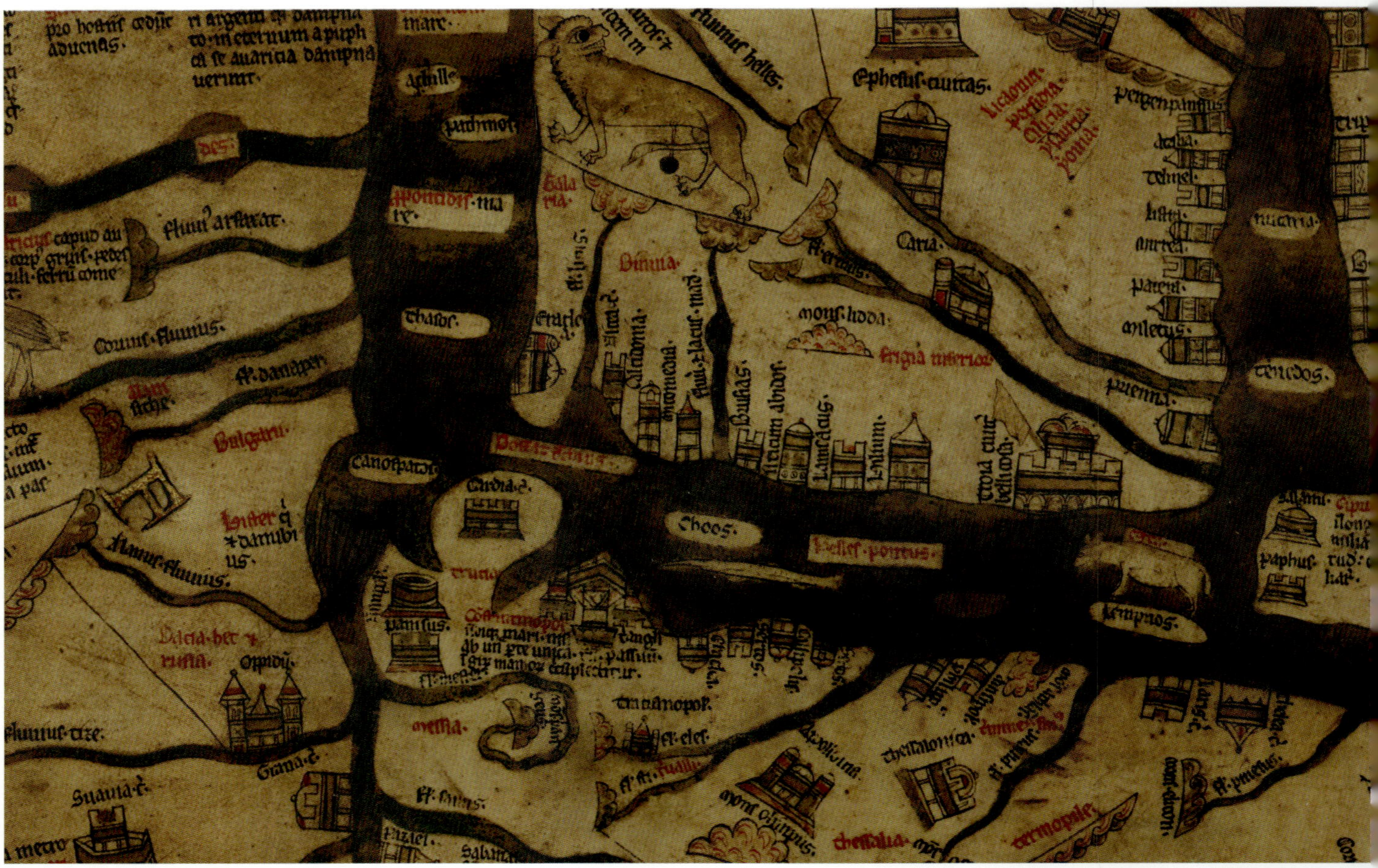
Ephesus ciuitas
Caria
mons Ida
Choos
Tenedos
Thasos
Patmos
Achille
Bulgarii
fluuius tize

has been lightened, but it has not been possible to restore the creamy white that must have been its original colour. While this means that the reproduction unavoidably lacks something of the vivid impact that the map must have had when it was made, it does much to show what this impact must have been and to make of the map once more a visually interesting object. In a very few cases lettering badly rubbed on the map has been rewritten.

Even without this enhancement the map is impressive in its size and complexity. It is drawn on a single, extraordinarily large piece of parchment, the whole skin of a calf. With the neck of the animal at the top, it is shaped like the gable end of a house, overall some 5 feet 2 inches high and 4 feet 4 inches wide: 1.59 metres from the apex to the bottom, 1.09 metres from the base of the triangular top to the bottom and from 1.29 to 1.34 metres across. Most of this space is taken up by the map itself in a vast circle; around it are the outer band naming the cardinal points and an inner band identifying the twelve winds of classical authority, the four major winds each shown by a tiny naked human figure with grotesque head, kneeling on all fours and puffing hard, the others by a dragon's head, blowing with open mouth.

Above the map itself is Christ in judgement, with the saved being led to paradise on his right hand and the doomed being led to hell on his left; below him the Virgin Mary, with bared breasts, pleads in four lines of French verse for those who sought her intercession – this and other inscriptions outside the map are given, with translation, in Appendix 1.[2] The world was God's creation, a point brought home by the letters MORS, the Latin word for death, spaced out beyond its bounds. Below the map on the left a seated man holds a sealed document of wholly thirteenth-century appearance; although he wears the triple tiara of a pope, the inscription on the seal shows that he is the Emperor Augustus, and he is giving the document to three named surveyors, ordering them to measure the world. An inscription in red round the edge of the whole parchment tells how one of them, Nichodoxus, measured the east, Policlitus the south, and Theodocus the north and west – inconsistently, it says this was by order of Julius Caesar. Below the map on the right a man on horseback, hand raised in salutation, passes a huntsman with bow and arrow, horn and a couple of hounds, who is apparently saying '*passe avant*' (go ahead)[3] – perhaps one of the Roman surveyors (he wears similar clothes) being given priority on the road, perhaps the medieval author of

The effect of restoring the map's colouring in an area centred on the Dardanelles. The Black Sea is on the left, Cyprus on the right

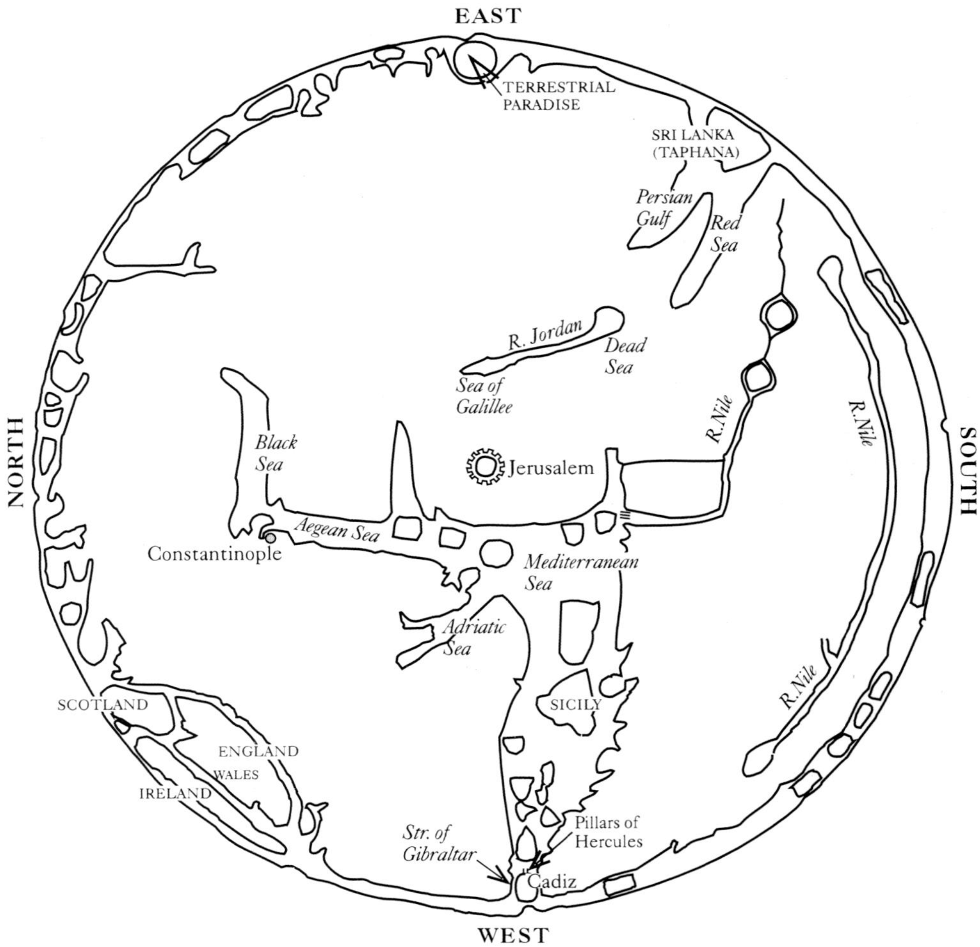

The layout of the Hereford map, a circle with east at the top, Jerusalem in the centre, was a conventionalised form of the way the 13th century envisaged the inhabited world

the map in person. It may, however, be a figure typifying the individual living in the confines of the world to which he is gesturing – or, as we shall see, there may be a topical or personal allusion that we can now only guess at.

The map itself looks strange to our eyes, though the places that western Europe knew through direct experience are in more or less correct relationship, while often much out of proportion and oddly skewed out of shape. Sea surrounds the whole land mass. East is at the top, and there, a walled island off the coast of Asia, is the earthly paradise with Adam and Eve taking the forbidden fruit; their expulsion is pictured on the adjacent mainland. At the foot of the map are the Straits of Gibraltar, with the Pillars of Hercules set on the island of Cadiz. To the left Spain, France and Italy form a single block of land; Italy is barely distinguished as a peninsula, though above it the inlet of the Adriatic is clearly shown. The Mediterranean is studded with islands – the triangular Sicily among them – and it stretches nearly to the centre of the map; at the sea's eastern end the Nile flows in from the south on either side of an island that represents

its delta. Opposite, past a circular peninsula that is the Peloponnese, a long channel leads northwards. This is the Aegean and beyond the straits at Constantinople it makes a right-angled turn eastwards as the Black Sea.

Closing the Mediterranean on the east is Palestine, flanked by two inlets of the sea, a long one to the north and a shorter one to the south. At

In the areas of – roughly – the ancient Roman Empire little is shown beyond rivers, mountains, provincial boundaries and towns, as we see in France (lower centre) and Italy (the broad peninsula further right)

The walls and three towers of Constantinople are pictured facing the sea, and thus upside down. To the left is the Danube delta, and beyond it the River Dnieper (fl' Danaper)

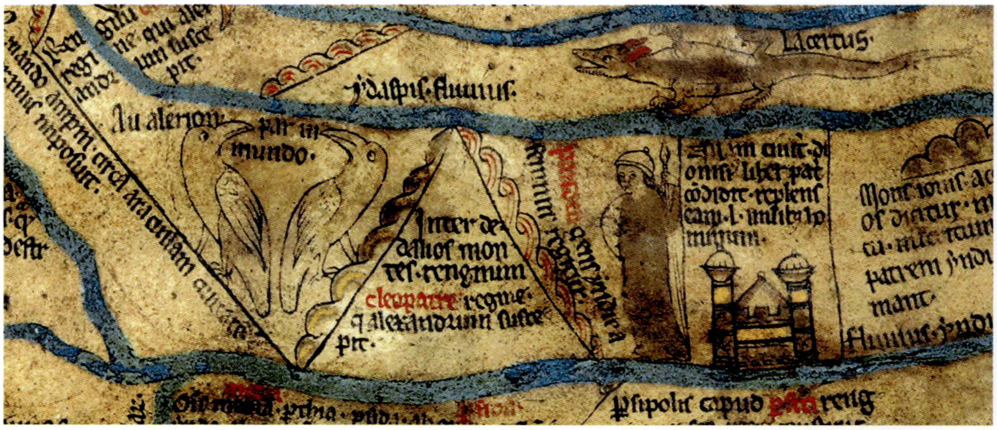

In India are, on the left, the avalerion *birds. There is only a single pair, which after sixty years produces two eggs; when they hatch the parents drown themselves. On the right is a woman warrior from a people governed by women*

the very centre of the map is Jerusalem, a circular walled city with the Crucifixion pictured above it, and further east is the River Jordan with the Dead Sea and Galilee. Beyond is the Euphrates with a massive structure that is the Tower of Babel, and to the east again are the Tigris and the Indus. India is the whole area beyond, with the Ganges, flowing eastward,

Samarcand is the city at the top, the Oxus the river on the left and the Bactrus the river at the bottom. Here the pelican revives its young with the blood from its own breast, and on the left are the birdlike people called the cicone

on the left. To the north of the Ganges and the Black Sea is the region called Scythia, which occupies much of the top-left quarter of the map; here we see the Caspian Sea in the form of a river flowing to the north. On the opposite side of the map are the Red Sea and the Persian Gulf and, disconcertingly, the island of Sri Lanka (Taphana) lies at their mouth. There are in fact islands all the way round the land mass, and by far the largest is England and Wales in the north-west, the bottom left – Scotland as well as Ireland is a separate island. Parallel to the outer coastline at the south of the map, a long river with a lake at each end effectively separates southern Africa from the rest of the continent. This is the upper Nile; though it does not visibly connect with the river that flows into the Mediterranean, it was taken to be the same river, flowing underground for part of its length. Rivers, though denser in Europe than elsewhere, are

Both the Red Sea and the Persian Gulf are coloured red. At their mouth is Taphana (Sri Lanka), where there are dragons

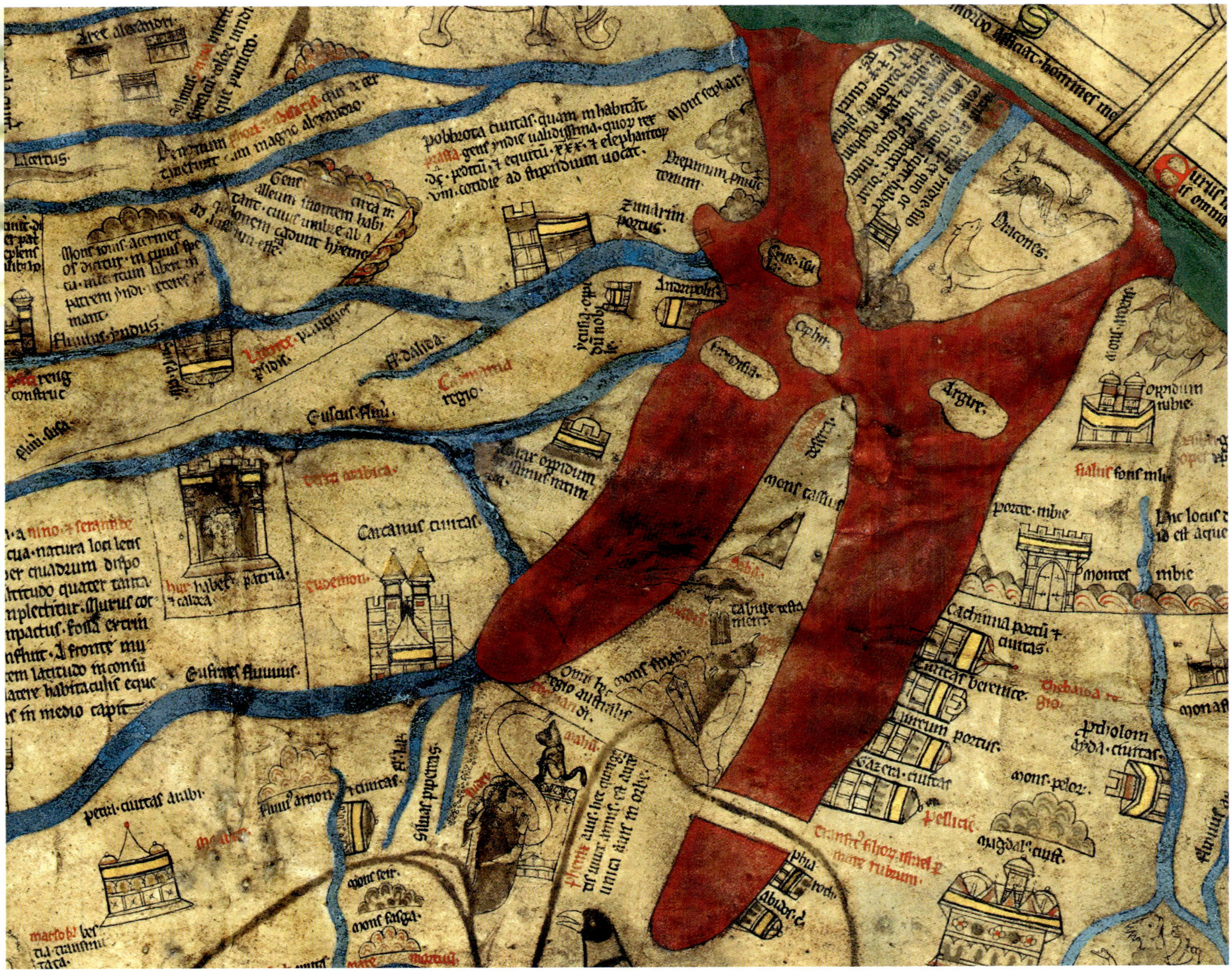

The essedones *in northern Asia eat their parents' corpses, thinking this better than leaving them to the worms. To the right is the golden fleece, from the Greek legend of the Argonauts*

shown throughout the map, and so are mountain chains, isolated peaks and boundary lines between peoples and provinces.

All the land on the map is covered with writing and pictures. However, west and south-east Europe, Asia Minor, Palestine and the north African coast differ from the rest – as we shall see, it may be significant that this is broadly the area of the old Roman Empire. Here it is mostly just the cities that are shown, conventionally pictured and named – indeed, they are so numerous that there is hardly room for anything else. Elsewhere there are a few towns, but otherwise the space is taken up with innumerable snippets of information, most of them accompanied by a picture. Some are simply geographic – the note on Cyprus gives its dimensions, 180 miles (290 km) long and 120 miles (193 km) wide. Some are biblical, historical or literary, recalling events that occurred in the past – it was in the mountains of Armenia that Noah's Ark came to rest, for instance, and not far away was the land where Jason found the Golden Fleece to take back to King Pelias. But more tell us what we would find in these distant places – strange ani-

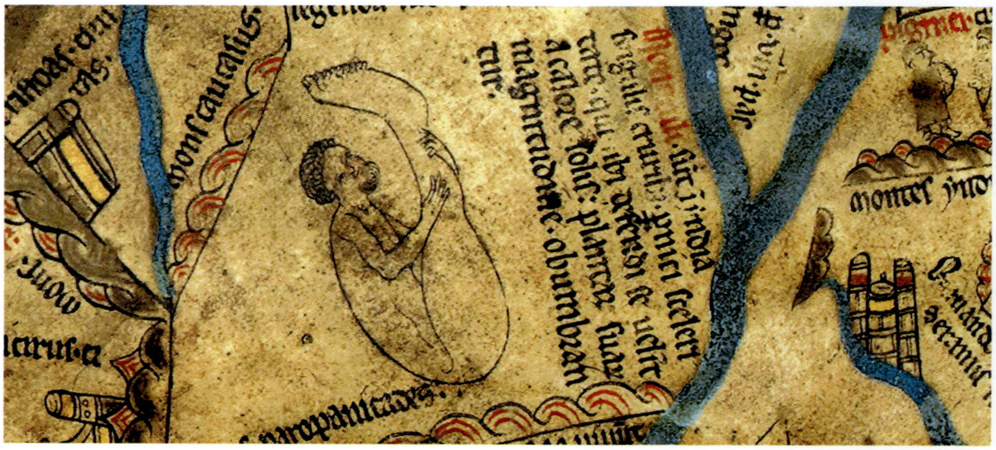

The monoculi *of India had only one leg, but they moved at great speed and their foot was so big that it could be used as a shade from the sun*

mals, birds and plants, people with weird customs, monstrous races of peculiar physique. The whole forms a kind of encyclopaedia, arranged geographically. On the map itself all the wording is in Latin – it is only a few of the inscriptions outside the border that are in French. French was at this time widely spoken by the English upper class, and its use here suggests that the map's author envisaged a readership beyond the learned and scholarly circles for whom Latin alone was appropriate.

The Construction of the Map[4]

In January 1999 the map was removed from its outer case – a complicated and lengthy operation – so that a close examination could be made of its handwriting by M. B. Parkes and of its graphic work by Nigel Morgan. At the same time Christopher Clarkson, as conservation consultant, made his first examination of the map's physical condition. Although the time available for all this work was severely limited, their reports transformed our knowledge of the map, not least our knowledge of when and how it was made.[5] It was not possible, however, for them to see the back of the map, as disturbing its system of support risks damaging the map and can be contemplated only in extreme necessity. No photograph or note was made of what – if anything – is on the back of the map when it was mounted on this support system in 1982 nor on any earlier occasion when conservation work was done, though we may reasonably suppose that any substantial markings or early notes would have been recorded.[6] But the map may still have surprises in store for us.

However, the name of the map-maker has long been known. Very few medieval maps reveal the name of the author, but the Hereford map is one of the exceptions. A note in French in the lower left corner asks all who hear, read or see the map, which it calls '*cest estorie*', this story or picture, to pray for Richard of Holdingham (Haldingham) or Sleaford (Lafford) who made it.[7] Holdingham lies just outside Sleaford, in Lincolnshire, and

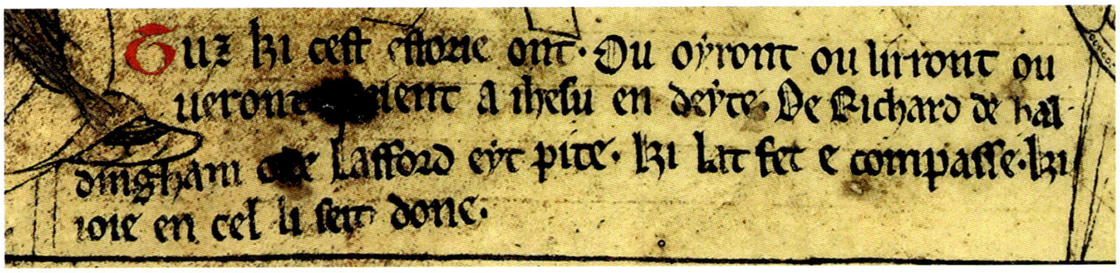

The inscription naming Richard of Holdingham or Sleaford as author of the map

The realistic detail at Lincoln shows local knowledge. Below are Northampton and Ely

the map itself shows a connection with Lincoln: the city is drawn more elaborately than any other in England, with the cathedral or castle on the hill and a house-lined street leading up from the River Witham. A less obvious association with Lincolnshire is provided by the precisely drawn circular labyrinth on Crete; its pattern exactly copies not only labyrinths laid out on the floors of Chartres Cathedral and some other large churches in France and Italy but also thirteenth-century labyrinths of turf in the open air at Alkborough in Lincolnshire and some other places, nearly all in neighbouring counties of eastern England.[8]

Turning to other sources, we find that one Richard of Battle (*de la Bataylle* in French, *de Bello* in Latin), in 1260 a rector in Kent, was a canon of

The pattern of the labyrinth on Crete is exactly that of one on the floor of Chartres Cathedral (below left) and of a turf maze at Alkborough in Lincolnshire (below right)

Lincoln by 1265 and held the prebend of Sleaford there by 1277; he died in 1278. He is not named Holdingham, but if Battle was his family name Holdingham may have been where he was born or lived; such an alternative surname would be normal at this time. Another Richard of Battle then appears. Our only household accounts of a medieval bishop of Hereford, in 1289–90, record gifts of venison to him and of money to his servant, and from 1305 until his death in 1326 he was a non-residentiary canon of Hereford and held the prebend of Norton in Herefordshire. Meanwhile, besides acquiring other benefices elsewhere, he had had a notable career in the diocese of Salisbury, becoming a canon of the cathedral in 1298, the bishop's vicar-general in 1299 and keeper of the spiritualities when the see was vacant in 1315.

That the first Richard of Battle died in 1278 is a recent discovery.[9] Earlier, some scholars had thought this was the career of one man, not two; thus Nigel Yates saw no insuperable objection to Richard of Battle being someone who lived to be at least eighty-six – not impossible even in the fourteenth century.[10] We see how well this would explain the map's early history, providing in its author the link between its Lincolnshire origin and its home at Hereford. We may still preserve this link by supposing the two men were closely related – perhaps uncle and nephew – and closely associated. However, the identification of Richard of Holdingham with the first Richard of Battle is entirely circumstantial. Yates warns us against taking it for granted and mentions another possible candidate, Richard of Sleaford, rector of Normanby-le-Wold, 17 miles (27 km) from Lincoln, from 1283 to 1294.[11] We can only say that it seems likely, but far from certain, that Richard of Battle, canon of Lincoln, was the Richard of Holdingham named on the map.

It has long been recognised that as the map shows both Conway and Caernarvon it could not have been drawn before work began on these two castles, in 1277 and 1283, while Michael Prestwich has suggested that its drawing of Berwick-upon-Tweed as a great fortress, the most prominent place in Scotland, points to a date after Edward I had captured it in 1296 and planned its rebuilding.[12] This fits in well with the conclusions of M. B. Parkes on the map's handwriting, which he dates between about 1290 and 1310, and of Nigel Morgan on its drawing: its curved form of the crucified Christ does not appear in England before the 1280s, and the top border's naturalistic leaf ornament, with winding scrolls, not until about 1285.[13] The map was probably drawn not long before or after 1300.

Clee Hill (Mons Cleoe) *may have been added from local knowledge. Below are Shrewsbury, Worcester and Hereford*

The map betrays its connection with Hereford as well as Lincoln. The name of the River Severn has been erased and rewritten to allow the entry of Hereford and the River Wye. These may be simply additions, but they may have been entered already, placed slightly differently. Either way they demonstrate local knowledge – but M. B. Parkes shows that they are in the same hand as the map's other inscriptions, an adjustment made in the course of its production.[14] Local knowledge, though misapplied, may also explain why the map gives the name of Clee Hill (Mons Cleoe), 25 miles (40 km) from Hereford, to the mountain shown as the source of both Severn and Dee. Clearly the map was associated with Hereford from the time it was being made: here at least its inscriptions were written.

If the map was produced at Hereford, perhaps twenty years after the first Richard of Battle had died, he may still have been Richard of Holdingham. The map is a book in graphic form, and just as a book might be copied long after its author's death, still naming its author, so too might the map; someone else will have brought the original up to date by adding probably Conway and certainly Caernarvon and Berwick, or indeed the map's scribe may have entered them on his own initiative, just as he made the local changes around Hereford. Richard may have designed – perhaps himself drew and wrote – a map kept at Lincoln that was then copied for Hereford; the original map may have been taken there – as a book might be borrowed for copying – or Richard's original sketches or drafts may have been sent as the basis for a new fair copy. Alternatively, the map may have been fully drawn at Lincoln and sent to Hereford to have its inscriptions and other lettering entered there from notes;[15] M. B. Parkes has suggested that the vertical crease down the centre of the map might have arisen from folding it by way of protection when it passed from artist to scribe.

The detailed reports by Nigel Morgan and M. B. Parkes now give us a clear picture of the successive stages of the map's production, and they are agreed that the graphic work was done first. On the map's surface many pinpricks can be seen and they may have played a part in transferring the design to the parchment; from the back of the map we could tell just how far these pinpricks form a meaningful pattern, and this is one of the items on Christopher Clarkson's list of what we should check whenever inspection of the back becomes possible.[16] Certainly, however, two holes were made by compasses to draw circles, one on Crete for the labyrinth, one at

Drafting lines for the planned course of the River Bactrus pass across the legs of the Bactrian camel; it was finally drawn at the camel's feet

Guide lines ruled above and below each line of writing are visible on the note marking the start of the route of the Exodus. The town encircled by the River Nile is Cairo, called Babilonia, *the new Babylon*

Jerusalem for the city wall and, being the map's centre, for the circles marking its outer edge. The coasts, rivers and mountains were outlined first, possibly with a plummet, a pencil of lead – some drafting lines are still visible – then the drawings of figures, animals, birds and historical sites.[17] These drawings were then outlined in ink, as well as some rivers, and a light colour wash added. At this point another artist probably took up the work; he inked over the geographical outlines and then entered the city symbols, perhaps drafting them first with plummet but possibly drawing them in ink from the start. Morgan suggests that as the cities and many of the animals are drawn in what was then an archaic style it can be argued that 'the artists certainly knew well an earlier map of the period before 1250'.[18] The final stage of the graphic work was to colour the rivers, the seas and the foliage drawn in the map's border.

Next the inscriptions were entered by a single scribe, who ruled lines for four of the longer ones but was otherwise guided only by the drawings and outlines.[19] He was a skilled craftsman: there are no signs of serious problems in fitting inscriptions into the available space, though a few are slightly constricted, and occasionally the spacing shows how one was entered before another, as on the boundary of Africa and Asia. Not only the names of provinces but also the names of animals and peoples were written in red, and where one of these names occurred in the middle of an inscription the scribe would estimate, mostly very accurately, how much space to leave to fill it in. He was, however, no scholar: some of the few mistakes arise from failing to understand what he was copying, and he expanded incorrectly some words that must have been abbreviated on the

The red note on either side of the mountain boundary between Asia and Africa was written before the black note identifying Alexander's camp, as its second line would otherwise have been further to the right

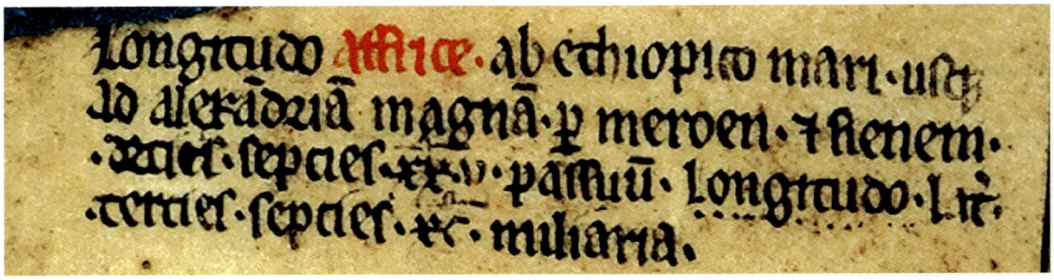

In the third line the map-maker accidentally wrote longitudo, *length, instead of* latitudo, *breadth. He thus put a row of dots below it – a normal way of cancelling a word – and added* lat'

exemplar.[20] He made a very few corrections. The changes around Hereford are one. Another is in the note on the length and breadth of Africa, both accidentally called '*longitudo*', length; the second has a dot below each letter – an accepted neat alternative to crossing out – and '*lat*'' (*latitudo*), breadth, added.[21] Finally another experienced craftsman entered the red lettering along the outer border and the letters in gold: the names of the continents and India and of the cardinal points, and 'MORS' outside the map. In doing this he wrote 'AFFRICA' across Europe, 'EUROPA' across Africa, 'a mistake whose dimensions inspire a certain awe'.[22]

In the highly ornamental lettering around the border one letter might be written within another. The word on the right is OCCIDENS, *west*

2

The Map and Its History

The Triptych[23]

The earliest known reference to the map is a note on Hereford Cathedral written by Thomas Dingley of Dilwyn, Herefordshire, in the early 1680s or a little earlier: 'Among other curiosity in this Library are an Map of ye World drawn on Vellum by a Monk Kept in a frame with two doors – with guilded and painted Letters and figures.'[24] This closely matches the opening words of the earliest reference in print, by Richard Gough a century later: 'In the library of Hereford cathedral is preserved a very curious map of the world, inclosed in a case with folding doors, on which are painted the Virgin and the Angel.'[25]

The case or frame that they mention was recorded in a coloured drawing by John Carter in about 1780.[26] It was a triptych, about 10 feet (over 3 metres) wide with its wings open. In the centre was the map, and above it

John Carter's drawing of c.1780 is our best evidence of the appearance of the complete triptych

The central oak panel is all that survives of the triptych that originally held the map

was a painted canopy, with carved oak-leaf crockets and a central finial; on the centre panel, as a narrow border around the map, scrolled decorations with dragon-like animals were painted in red on a white background. On the wings was painted the Annunciation, the Virgin on the right wing, the angel on the left holding a scroll inscribed 'AVE MARIA', both dressed in red with blue undertunics; it is not known what was on the back of the wings, visible when they were closed as doors covering the central panel with the map.

The only part of the triptych that survives is the gable-shaped central panel, comprising six oak planks within a moulded frame, with oak-leaf

This indentation in the centre of the triptych panel may have been made by the compass that drew both the outer circles of the map and, at its centre, the circular wall of Jerusalem

mouldings along the two sides at the top. In 1989 radiocarbon dating of two samples from the planks and one from the crockets showed that the wood was medieval. Later, in 2004, analysis of the tree-rings gave a more precise and consistent dating, showing that the oaks grew in Herefordshire, Monmouthshire or Gloucestershire and were felled between 1289 and 1311.[27] In 1855, when the map was undergoing restoration at the British Museum, Sir Frederic Madden, Keeper of Manuscripts, described the narrow painted border on each side of the map, then still visible, as 'the dragon-shaped animal that so often appears in MSS. executed about 1300', which 'at once fixed the date of the frame itself'.[28]

It is reasonably certain then that the triptych dated from the same time as the map and even if it was unconnected with the map it would still be of great interest: it is the earliest triptych of substantial size from England and few thirteenth-century English panel paintings of figures are known.[29] However, although we should not take this for granted, it seems most likely that the triptych was in fact constructed to house the map. In the nineteenth century the map was fixed in its case by a thin brass strip along each side, secured by a large number of round-headed brass nails which must have passed through the extreme edge of the parchment,[30] and this arrangement probably dated from the Middle Ages, possibly from the time the map was first set in the triptych. Indeed, Christopher Clarkson suggests that the parchment was already attached to the panel when the map was drawn; a small conical depression in the wood can be interpreted as made by the compass point to draw the circles of Jerusalem and the map's frame.[31]

From the late eighteenth century onwards the triptych was gradually dismembered. The two wings were taken off, probably soon after 1780, and seem to have been destroyed or lost, though the intriguing possibility remains that they survive somewhere, unrecognised.[32] Sir Frederic Madden had scarcely seen and dated the back panel's decorated border, which was painted on top of whitewash, before the paint was removed – at the British Museum, in the course of the map's conservation in 1855. In his diary Madden tells how the Museum's binder

> . . . said that in taking up the vellum and cleaning the brass nails, the whitewash had scaled off, so that it was thought best to take the whole away! I am angry at such folly, but the deed is done, and I suspect that no one but myself will ever notice the deficiency.[33]

As far as we know, for nearly a century and a half no one did.

In 1862 plate glass was ordered to be put in front of the map, and in

The earliest photograph of the map, 1868, shows the brass strips and round-headed nails that held the map to the oak panel

1868 new doors were added, probably replacing the curtain that had hung in front of it. The finial at the top of the case disappeared between 1931 and 1946, probably when the map was evacuated during the Second World War. Finally, following the map's further restoration – which is probably when the brass strips and nails were lost – a new case was provided in 1948 and the old one discarded.[34] In 1989 Martin Bailey, who was working for *The Observer* newspaper, saw John Carter's drawing of the triptych and, recognising its importance, wondered what had happened to it. He contacted the Chapter Clerk at Hereford and together they went to the most likely place where it might still be: a storeroom that had been the cathedral's stable. There,

> ... after shifting through considerable debris, we found a large wooden panel leaning against a wall, covered with a thick layer of dust. It had a decorated Gothic gable and on the front was the unmistakable outline of the mappamundi, picked out in nail holes. It was clearly the central panel depicted in the Carter drawing ...

The radiocarbon dating that established its medieval date was carried out at the expense of *The Observer*, and it was placed on exhibition with the map.[35]

The Purpose of the Map

How the map came to be drawn for Hereford, what purpose it was meant to serve there, are questions that have been much discussed since 1989 when the map came into prominence and the triptych was discovered. As we shall see, similar large world maps were painted on the walls of royal residences, probably in the context of other paintings of allegorical and moral import. Of themselves they could be viewed as a form of enlightenment or diversion for the literate, who could read them, or for the illiterate, to whom they could be explained, conveying interesting, even entertaining, information in a learned and pious setting. Rather more they will have come to be objects of contemplation; they will have been viewed as a basis for meditation focused on the world God had created.[36] This is probably how the map was seen at Hereford, and perhaps we need look no further to explain why the map was drawn. The bishop of Hereford from 1283 to 1317 was Richard Swinfield; he was a man of wide experience, and the

Dictionary of National Biography described him as 'a bountiful patron of learning'. Among other posts earlier in his career he had been chancellor of Lincoln Cathedral. It is perfectly possible that, having seen Richard of Holdingham's original map at Lincoln, he commissioned a copy for Hereford simply to be an instructive and uplifting adornment of the cathedral. Alternatively we may see the second Richard of Battle as the link between Lincoln, where he had taken holy orders, and Hereford, where he received gifts and preferment from Swinfield; if he was in fact a relative of Richard of Holdingham we might suppose a particular interest in the map – or particular access to its drafts – that might lead him to have the copy drawn as a gift to Hereford.[37]

If the parchment was already attached to the central panel of the triptych when the map was drawn, or if the triptych was made for the map very soon after it came to Hereford, the possibility arises that it was used from the start as an altarpiece, a representation of Christ in judgement above the whole of God's earthly creation – in greater detail than any worshipper could see, but this was hardly the point. This was the understanding of Richard Gough in the late eighteenth century and was the view of some – but by no means all – of those who wrote about the map in the next two hundred years.[38] In 1989 the question again came to the fore, and in 1994 Marcia Kupfer seriously questioned this view of the map's role: she urged the possibility that the map was placed in the triptych long after both were made, but she argued that in any case the map was quite inappropriate for an altar. Earlier, in the eleventh and early twelfth centuries, some monastic schools used a world map in teaching; it would seem less well suited to the more rigorous, formally structured education of the late thirteenth century, but one possibility is that the map might have been used at Hereford for less formal instruction, as an aid not to teaching but to preaching.[39]

In 1998 Valerie Flint introduced a completely new approach to the origin and purpose of the map. In an important article she argued that we should see the map in the context of contemporary events at Hereford and set herself the task of explaining the enigmatic scenes in its lower corners. For nearly twenty years Richard Swinfield had been closely associated with his predecessor, Thomas Cantilupe, who had been bishop of Hereford from 1275 until his death in 1282; he owed much to him and clearly held him in great affection and respect. Throughout the thirty-four years when he was himself bishop, Swinfield worked hard to secure Cantilupe's canonisation as a saint; he succeeded, but not until 1320, three years after he had himself died. Valerie Flint suggested that the pictures below the map refer to some of Cantilupe's achievements. The Emperor Augustus, wearing the pope's triple tiara, should be seen also as the pope, and the three surveyors as the three supervising collectors of a tax imposed by a church council at Lyons in 1274 that Cantilupe attended. The scene opposite recalls how Cantilupe in 1277–8 won back for the bishopric hunting rights in the Malvern Hills that the earls of Gloucester had usurped; the huntsman is one of the earl's foresters, while the rider, allowed to pass, may be Cantilupe himself.[40]

Much of Valerie Flint's interpretation may be questioned. She herself showed – for the first time – that the first Richard of Battle died in 1278 but argued that he made the map himself at Lincoln, leaving it unfinished; it was completed, with the addition of the scenes below, and taken to Hereford by Swinfield's protégé the second Richard of Battle in 1282–3 when Swinfield became bishop. This cannot now be accepted as we know this was not how the map was made, and the proposed date is stretching the parameters argued by Nigel Morgan and M. B. Parkes to their very limits and beyond: as we have seen, they both, for quite different reasons, give their preferred date for the map as between about 1290 and about 1310.[41] Again, Nigel Morgan argues that the rider in the picture could not possibly be Cantilupe, as no artist in the thirteenth or fourteenth century would ever show a bishop without a mitre.[42] But what is important in Valerie Flint's work is that she has seen that the scenes below the map may well be significant, she has faced up to the problems they present and she has pointed clearly to the possible relevance of topical events at Hereford, above all Swinfield's efforts to secure Cantilupe's canonisation. Even if we now see the map as a copy of Richard of Holdingham's original work it need not have reproduced its exemplar or draft in every detail; we have seen how local knowledge at Hereford was introduced into the map itself, while the borders would be seen as free space – outside the book's text, so to speak – and could well be used to convey a local topical message.

In 2004 Dan Terkla took up the map's association with Cantilupe proposed by Valerie Flint, though without reference to the pictures in the border, and suggested that the map dates from 1287, when the potential saint's body was moved to a new tomb; this was to be the shrine that was the centre of his cult, the goal of the pilgrims who were already flocking to

The tomb of Thomas Cantilupe, late 13th century, shown here in an engraving of 1831, in the north transept of Hereford Cathedral

Hereford to seek his intervention. The tomb, shorn of its medieval adornment, survives close to the east wall of the cathedral's north transept; formerly it abutted directly on to the wall. Between the tomb and the end of the wall, and centred below a tripartite window, is a horizontal series of eight stone blocks inserted into rectangular holes in the wall; these can be seen as either the remains of a line of stone brackets cut flush or as filling the spaces left by a line of wooden brackets. Their length and their height from the ground mean that if the triptych containing the map had rested on these brackets it would have fitted just below the window while its

Eight holes in the wall of the north transept, now filled with stone blocks, may have been made for brackets to support the triptych and map

Reconstruction of the original site of the map, as proposed by Dan Terkla, next to Cantilupe's tomb in the north transept

doors, when open, would have projected nearly 11 inches (27 cm) beyond the end supports, an acceptable overhang. Jerusalem, with the Crucifixion, and the scene of the Last Judgement would be at a height to be clearly visible to the pilgrims visiting Cantilupe's shrine; it would be a part of the pilgrimage experience. Terkla suggests that the second Richard of Battle may have presented the map to the cathedral, as a gesture to his patron, Swinfield, in 1287 when the new tomb was inaugurated.[43]

More recently Thomas de Wesselow has argued that there are detailed objections to Terkla's placing of the map beside Cantilupe's tomb, and that what nineteenth-century guidebooks to the cathedral have to say of its location should be taken seriously; they explain that the map had originally hung from iron clasps on a pier in the south chancel aisle. These clasps, like large flat hooks, can be seen in an engraving of 1819, and de Wesselow suggests that they fitted into two horizontal battens that are nailed to the back of the triptych's central panel. In this position the map would still have been designed to play a part in the pilgrimage to Cantilupe's shrine and, indeed, de Wesselow sees relevant allusions in the map, among them the mitred figure leading the procession of the saved at the Last Judgement, the rider in the border on the pilgrimage through life and the three surveyors sent out by the papally attired Augustus, who may be seen as the three commissioners that the pope sent in 1306 to enquire into the life and miracles of Cantilupe. He too suggests that the second Richard of Battle commissioned the map, and connects it with remodelling that part of the cathedral in 1305–7.[44]

Iron clasps in the wall of the south choir aisle of Hereford Cathedral, seen in an engraving of 1819, may mark the original location of the map

There has been so much ingenious, learned and interesting speculation about the origin and purpose of the Hereford world map that it may be helpful to summarise what we actually know. It is likely, but not certain, that Richard of Holdingham who is named on the map was Richard of Battle, canon of Lincoln and prebendary of Sleaford, who died in 1278. It is likely, but not certain, that the second Richard of Battle, canon of Hereford and prebendary of Norton, was related to him. The map was originally designed at Lincoln, before 1278 if Richard of Holdingham was Richard of Battle, but the map as we have it was produced at some point between about 1290 and about 1310, probably (because of the prominence of Berwick) after 1296; it thus dates from the time when Richard Swinfield was bishop of Hereford. If we accept that the indentation in the triptych's panel was made by the compass that drew the wall of Jerusalem and the map's outer circles and accept too that the wood of this panel came from the Hereford region, then the artists' work that was the first stage in the map's creation was carried out at Hereford; certainly it was there that the map was completed with the writing of its inscriptions. There is no evidence to connect the map with Thomas Cantilupe and his nascent cult; the scenes below the map may have quite different connotations. The stone inserts beside Cantilupe's tomb in the cathedral, the iron clasps in the south chancel aisle, may point to the map's early location or locations but may have nothing to do with the map at all. To say this is not to devalue the work of Valerie Flint, Dan Terkla and Thomas de Wesselow – their arguments are reasonable and full of interest and may well contain the truth. But we must acknowledge that we have very little hard evidence how, when or where the map was made, none at all why; the most that can be offered is informed – often very well-informed – guesswork.

THE MAP'S SUBSEQUENT HISTORY

The Map's Subsequent History

It is not only in general ageing and in the loss of some of its colour that the map has suffered over the years. We have seen that at some point it was folded lengthways, perhaps while it was being made, perhaps later. In the lower left corner there is heavy scoring, centred on France but extending into adjacent areas; this has been explained as a medieval or later expression of anti-French feeling,[45] and from the form of the scoring it is likely to have been deliberate, not accidental damage – though we may note that

Heavy scoring across France on the map seems to be centred on Paris, a huge multi-towered city, perhaps resulting from anti-French feeling in the later Middle Ages

'a quantity of glass lanthorns' was found 'piled against the map' in 1813.[46] There are at least seven holes in the parchment; the most obvious are one beside the 'M' of 'MORS', one a little to the south of the island of paradise, one in the Red Sea and one beside the south-by-south-east wind, from which a crack extends some way along the southern coast of Africa. These holes have all been repaired with parchment patches, painted over, as appropriate, to avoid disfiguring the map.[47] Just below the last of these holes the inscription beside the horned and beaked satyr has been mostly erased, perhaps deliberately.[48] Finally, right round the map its extreme edge is frayed and has been backed with new parchment, damage that can

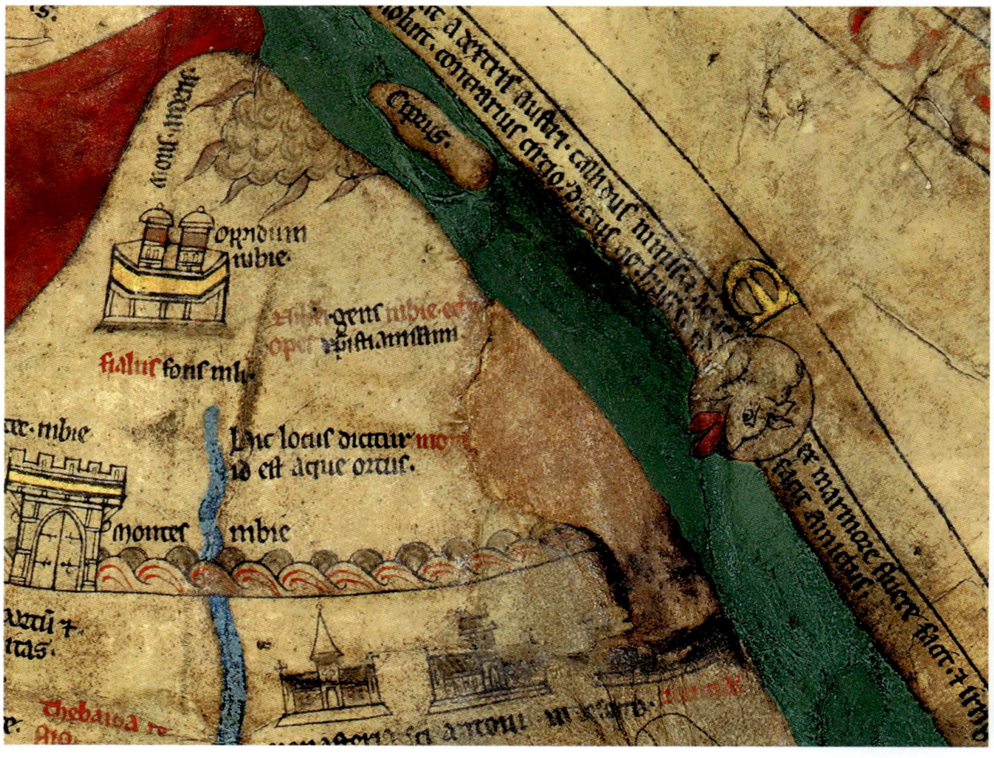

One of several holes repaired with patches of parchment, coloured so as not to disfigure the map. Between the patch and the Red Sea, on the left, is the burning mountain, mons ardens, *with flames issuing from each side*

The inscription erased beside the satyr in upper Egypt has so far eluded attempts to read it. It was in the desert here that fauns and satyrs tempted St Antony, founder of monasticism. On the left, on an island in the Nile, is a man riding a crocodile

be easily explained by the brass strips and nails, probably several times replaced, that held the map to the triptych.[49]

Whatever purpose the map was meant to serve and wherever in the cathedral it was placed, the Reformation and new geographical knowledge will have deprived it of much of its meaning. It would be seen only as an archaic curiosity and was probably now removed to the cathedral's library, housed in the Lady Chapel. It may well have been there in the early seventeenth century, when the signature of 'John Nicolles' was written beside the horseman in the lower right corner; he was perhaps a son or other relative of William Nichols, vicar-choral who died in 1635.[50] It was certainly in the library in about 1680 when, as we have seen, Thomas Dingley saw it and made the first surviving note of its existence. It was believed at Hereford in the nineteenth century that at some point the map had been hidden under the floor of the cathedral's Audley Chapel, a chantry beside the

The signature 'John Nicolles' in an inexpert hand of the early 17th century

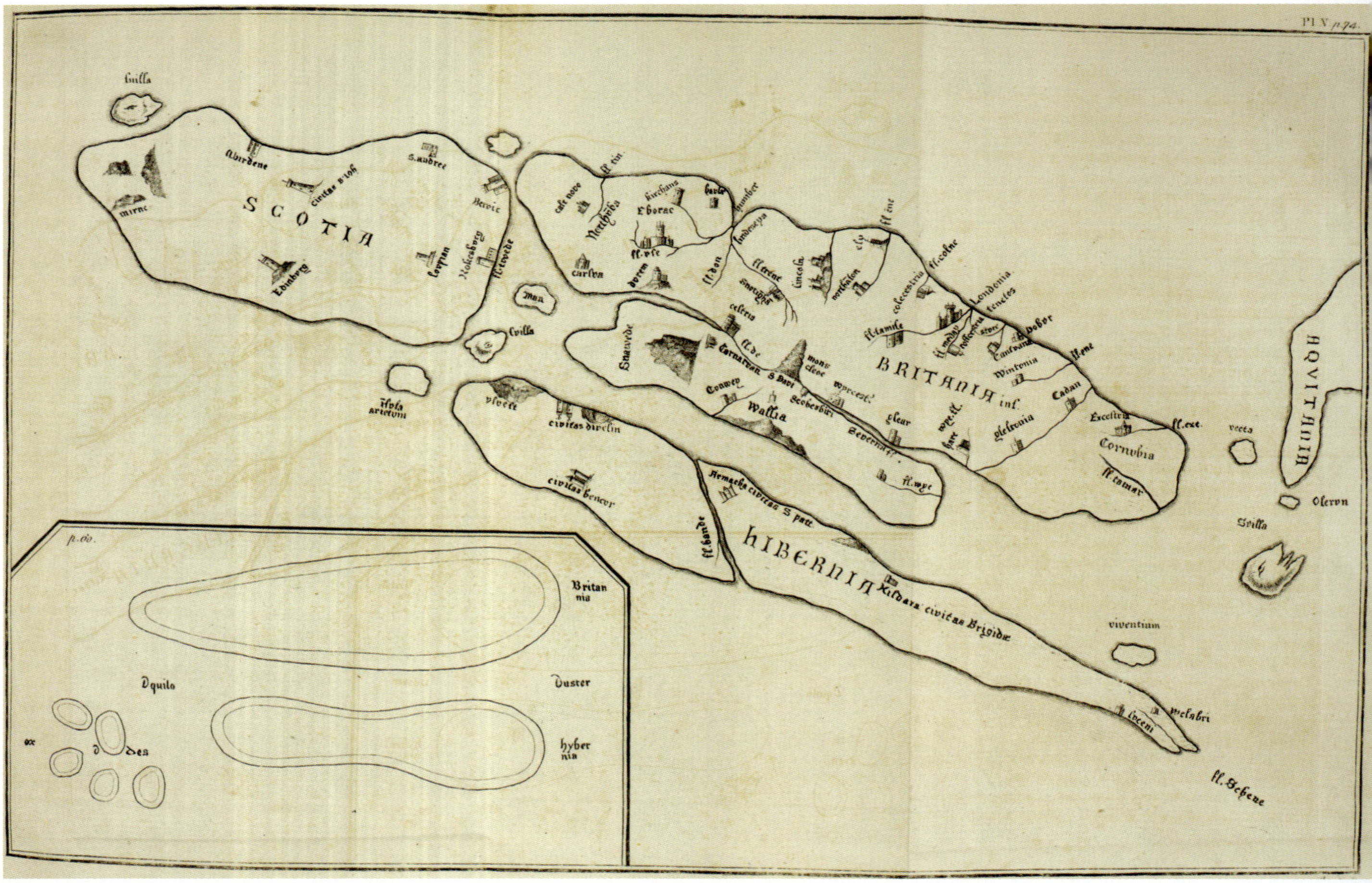

Engraving of the British Isles in Richard Gough's British Topography, *1780, the first printed reproduction from the Hereford map. Inset, bottom left, is a map of the British Isles from a copy of the description of Ireland by Gerald of Wales*

Lady Chapel; though there is no other evidence, this is quite possible – the Lady Chapel for a long time housed the library.[51]

We have seen that it was Richard Gough in 1780 who first referred to the map in print; in his two-volume work *British Topography* he introduced the map to the antiquarian public with a full description together with an engraving of its representation of the British Isles.[52] In the same year John Carter showed it, in the triptych, on the engraved title page of his *Specimens of the Ancient Sculpture and Paintings*.[53] Even so, little importance was attached to the map at Hereford; early nineteenth-century writers say that it had been found 'under a pile of lumber' and we have seen how lanterns had been piled against it in 1813.[54] In 1830 it was sent to London to be shown to the Royal Geographical Society; on its return to Hereford Thomas Ballard was commissioned to make a copy, completed in 1831, that is still in the Society's library.[55] From this another copy was made in 1841 for the Bibliothèque du Roi in Paris and this in turn was reproduced a year later by E.-F. Jomard in his *Les Monuments de la géographie*.[56] By this time it had been several times described and discussed by writers on medieval geography and maps.

THE MAP'S SUBSEQUENT HISTORY

In 1855 it was sent to the British Museum to be cleaned and repaired – when, as we have seen, the paintwork on the panel was lost – and since then, in the words of Bevan and Phillott, 'it has been treated with the most reverential care'.[57] It had been moved from the library to the vestry, also known as the treasury, by 1830; to a room above the north transept on its return from the British Museum in 1855; to the south chancel aisle – perhaps a return to its original location – by late in 1862; and to the south transept in the early twentieth century.[58]

In 1867 a project was set on foot in Hereford to publish by subscription a new and improved facsimile of the map.[59] This had three important results. First the map was in 1868, for the first time, photographed by Thomas Ladmore, professional photographer of Hereford.[60] Second, the facsimile itself was published in 1872 after delays caused by the Franco-Prussian War; it was prepared in Bruges by a lithographer who also had a workshop in Paris, and one of those working on the map was killed at the battle of Sedan. Third, in 1873, a book was published to accompany the facsimile, written with verve and with great learning by W. L. Bevan, vicar of Hay-on-Wye, and H. W. Phillott, rector of Staunton-on-Wye; they describe the map in detail, identify the sources of its information and relate it to other medieval maps – a pioneer work of great importance. Their work was followed up in 1896 by the German map historian Konrad Miller; one of his six volumes on the earliest world maps is devoted to a transcription and analysis of its inscriptions, the fullest account of the map until Scott D. Westrem's vastly more detailed work was published in 2001.[61]

The map was shown at an Art Treasures of England exhibition at Manchester in 1857, but it does not seem to have left Hereford again until the Second World War, when it was moved, for safety, possibly first to the wine cellar of Hampton Court, Herefordshire, then to a coal-mine at Bradford-on-Avon, Wiltshire.[62] After the war, in 1948, the map again underwent conservation work at the British Museum, and on its return to Hereford it was displayed in a new case, in the north chancel aisle, provided at the expense of the Royal Geographical Society; the case was made of wood from City of London churches that had been bombed in the Second World War.[63] In the same year the Royal Geographical Society published a brief account of the map by G. R. Crone, the Society's librarian, and in 1954 the first full-size reproduction, an uncoloured collotype on nine sheets, with a longer memoir by Crone.[64] Finally, in 1982, further work by the distinguished private conservator S. M. Cockerell at Grantchester brought the map to its present state of conservation, supported and held on teak boards. It returned to Hereford in a van, wrapped in Cockerell's own eiderdown 'so as to buffer against sudden changes in temperature'; 'The eiderdown which you so kindly lent', the Cathedral Librarian wrote a day or so later, 'has been dispatched today by parcel post.'[65]

In 1987–8 the map was exhibited at the Royal Academy, London, at the important Age of Chivalry exhibition of thirteenth- and fourteenth-century English art.[66] In November 1988, it was announced that the Dean and Chapter had instructed Messrs Sotheby and Co. to auction the map the following June; it was hoped to raise a sum that would go far towards meeting the cathedral's long-term requirements for running costs. This aroused public protest and controversy. Clearly the map was better known, more highly regarded than had reasonably been supposed. But besides concern for the map itself, the proposal raised questions of the legal protection of the national heritage, the maintenance of ancient church buildings and the role of the Church of England as custodian of monuments and treasures, questions of wide interest. The sale was postponed, and in 1989–90 the map was displayed in London at the British Library. In 1990 the Mappa Mundi Trust was set up, with a capital endowment from the National Heritage Memorial Fund, to take over the ownership and care of the map and of the chained library and other historic books at Hereford. A generous gift from Mr J. Paul Getty Jnr, KBE, enabled the Trust to put up a new building for their preservation and display.[67] This building, designed by Whitfield Partners, was opened in 1996.

In June 1999 an international conference was held at Hereford on the map and related topics, but it was not until 2006 that its proceedings were published; we have already seen how important some of the contributions have been to our knowledge of the map. Meanwhile, two books published in 2001 further advanced our understanding: besides Scott D. Westrem's definitive work, Dr Naomi Reed Kline set the map's texts and images in the context of medieval thought.[68] Debate continues on the map's origin and purpose. There is still much we should like to learn about the Hereford map.

3
The Map and Its Relatives

The Roman Ancestors

How important is the map? There are, after all, over a thousand medieval world maps in existence – they are by far the most common type of map that survives from the Middle Ages, and probably by far the most common type that was produced. However, most of these maps were of the simplest kind, no more than diagrams of the main regions of the world. Some were zonal maps, a circle divided into horizontal bands representing the tropical, temperate and frigid zones. Others made a gesture to geographical outline, the so-called T-O maps – a circle divided by a T (sometimes a Y) to show the three continents, Asia, Africa and Europe. These might seem two analogous ways of dividing the world into regions, the climatic and the geographic, but in fact each starts from a different basis. The T-O map covers only the inhabited world, whereas the zonal map is the entire globe seen from one side.[69]

We should not be misled by the Hereford map into supposing that well-educated persons in the thirteenth century thought of the world as a flat disc with Jerusalem at its centre. That this is how the Middle Ages saw the world is a modern myth – one that the Hereford map may have helped to foster.

Two common medieval diagrams of the world, a zonal map (left) and a T-O map (right)

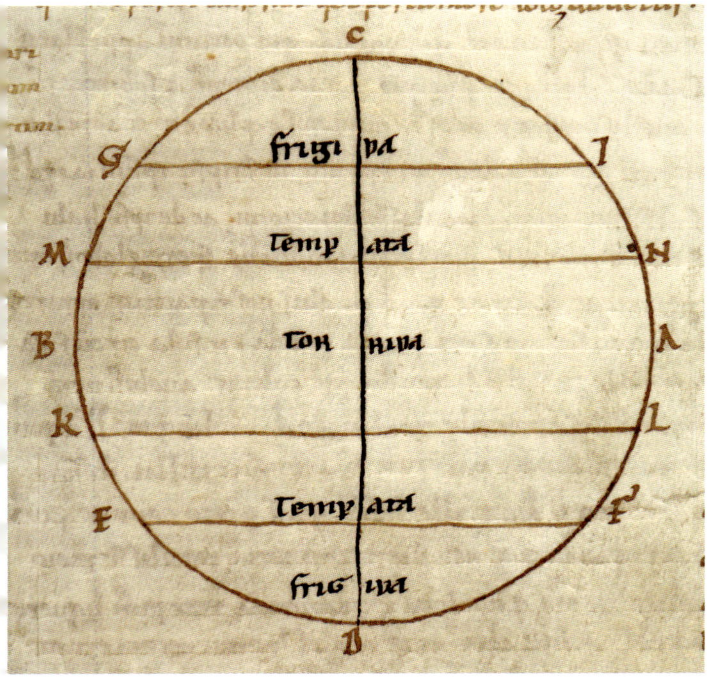

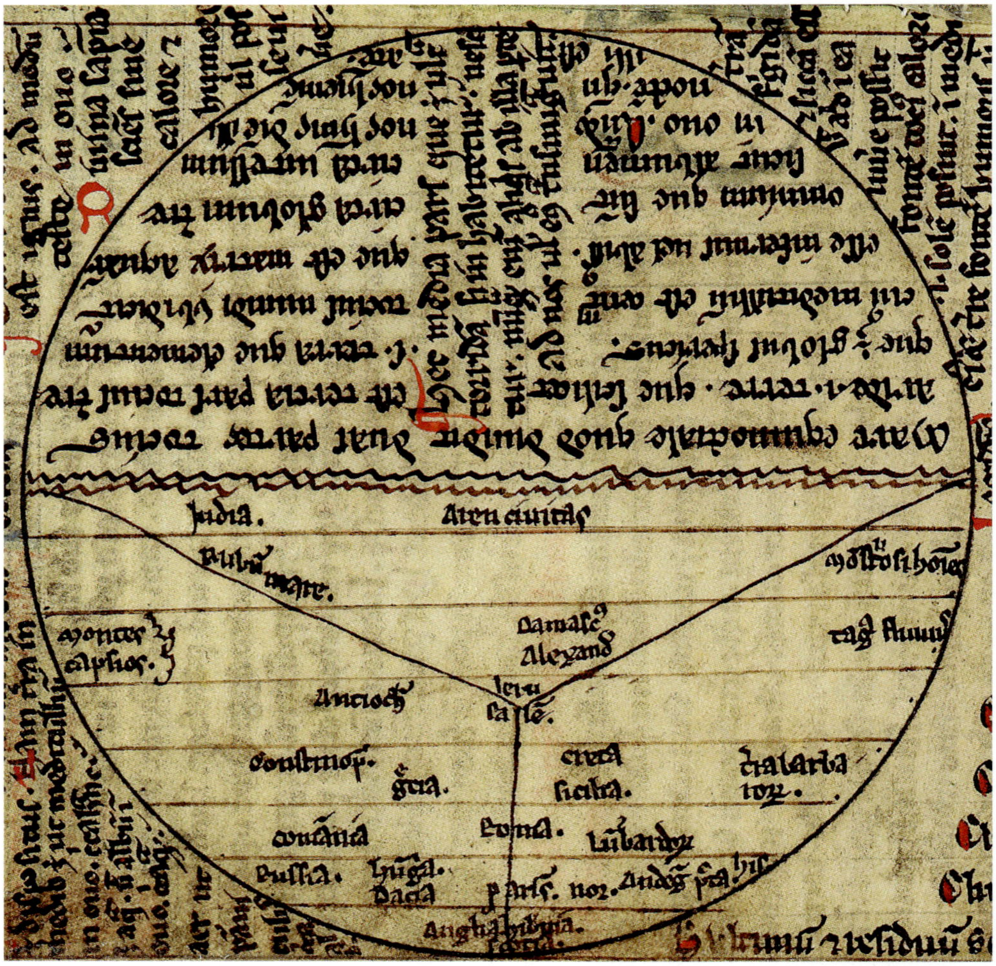

Map of the world seen from the side with south at the top, 13th century. In the unknown southern hemisphere are geographical notes. In the northern hemisphere horizontal lines mark climatic zones, and superimposed are a few place-names and the division into three continents. Asia, not Africa, is shown in the south, and some names are placed in the wrong continent

Contemporary scholars in fact envisaged the world as a sphere just as we do, but a sphere rather smaller than its actual size, with its known inhabited part in the north, separated impassably from its unknown southern half by a belt of ocean and by the unbearable heat of the tropics. The circle of the T-O maps, the ocean-bounded circle of the Hereford map, would be viewed as a conventionalised form – a kind of projection – of what really occupied perhaps rather more than half the surface of a dome, the northern hemisphere of the world. The southern parts of Africa that we see on the Hereford map should be understood as lying within this northern hemisphere – Africa was not thought to extend to the equator, let alone beyond.[70]

While most of the world maps that survive from the Middle Ages are schematic T-O or zonal maps, some maps of the inhabited world are more elaborate and detailed. Most illustrate books that have geographical content or implications – encyclopaedias, lists of words and places, biblical commentaries. Sometimes the map appears in so many manuscript copies of a book that we can accept it as an integral part of it – it was either

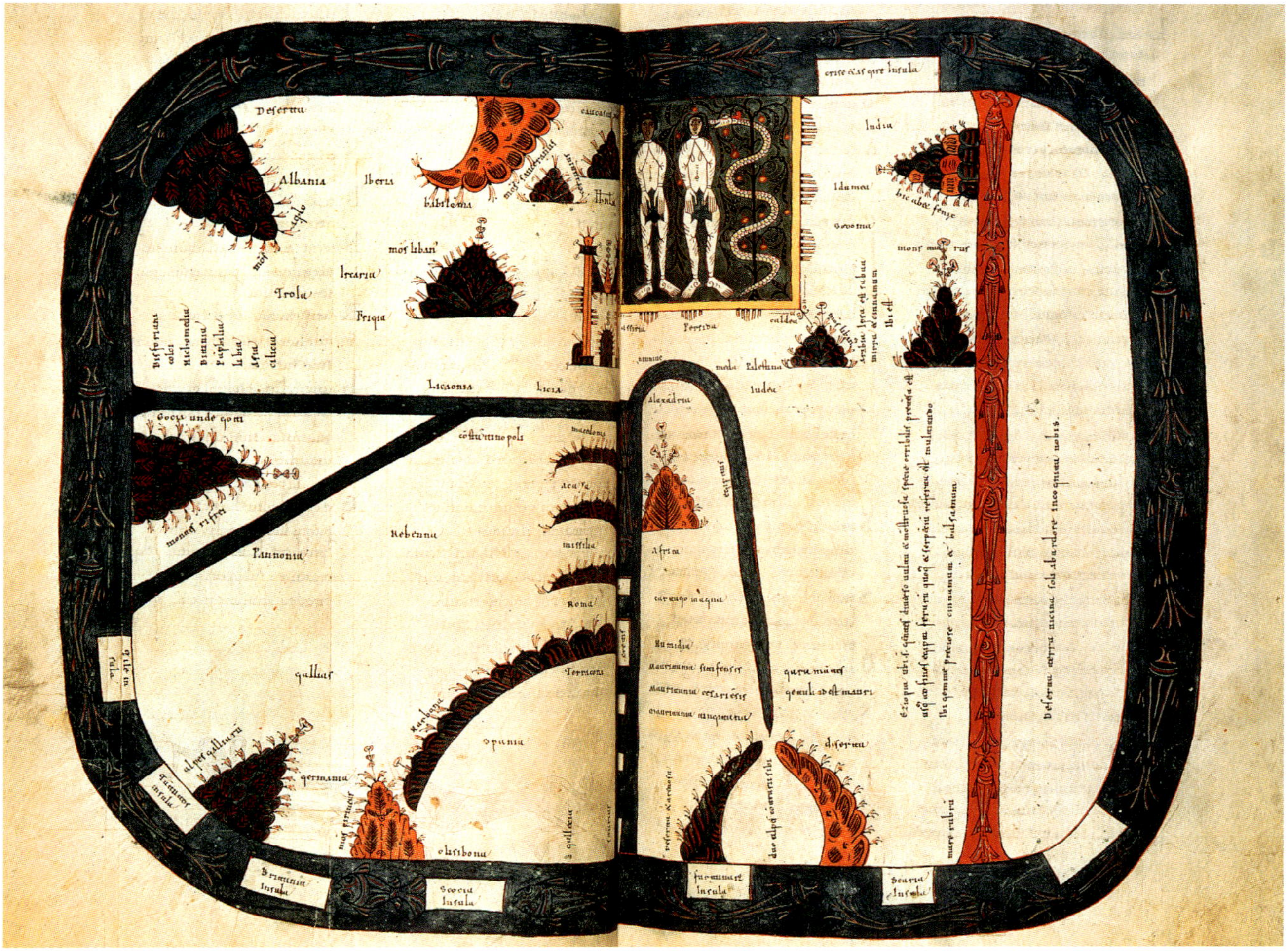

World map in an early 12th-century copy of Beatus of Liébana's commentary on the Apocalypse. East is at the top, with Adam, Eve and the serpent, the vertical brown line in the centre is the Mediterranean, and on the right, beyond the red band of the equatorial ocean, is the fourth continent of the southern hemisphere

drawn by the author when he wrote the text or added by a copyist at so early a stage in the sequence of copying that it became effectively incorporated into the work. Thus a map of the world appears in fourteen surviving manuscripts of the commentary by Beatus of Liébana on the Apocalypse, manuscripts that date from the tenth century to the thirteenth; most of them are rectangular in shape, distinctively Spanish in artistic style and brightly coloured, and they must derive from a common original which may well have illustrated the book when the author wrote it in the late eighth century. Like a few other medieval maps, all but one of these maps show the fourth continent that some thought might lie in the southern hemisphere, beyond the equatorial ocean.[71]

Another book consistently illustrated with a world map is the encyclopaedia by Lambert of Saint-Omer, and we know that this was the author's intention, for it was included in his autograph text, written in the early twelfth century. But an individual copyist might, on his own initiative and using some other source, change the map in the text he was copying or add a map to a text hitherto unillustrated. The encyclopaedic works of Isidore of Seville, written in the early seventh century, normally contain a more or less simple T-O map, but one twelfth-century copy now at Munich has a map showing cities, rivers and mountains that is visibly akin to the Hereford map, though only just over 10 inches (26 cm) across. We must be careful not to assume that the original author of a work had anything to do with a map that appears only in a copy made centuries later.[72]

Simple or elaborate, large or small, all these maps of the inhabited world are related to each other and to the Hereford map. In the broadest terms, features of their outlines are common to all. They all belong to a single, much ramified tradition which must go back to the Roman period.

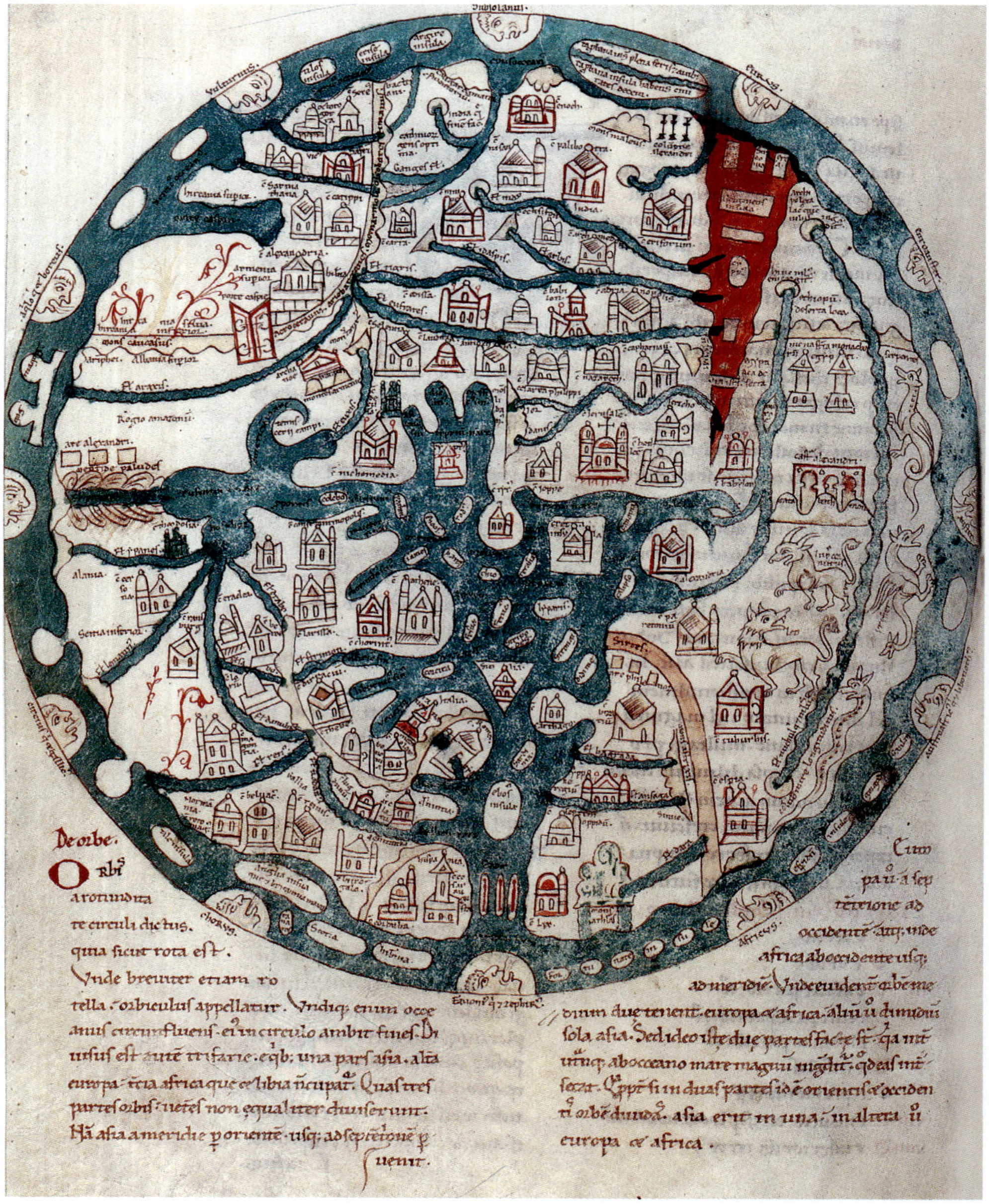

World map in a 12th-century manuscript of works by Isidore of Seville, drawn in northern France and now at Munich. The Hereford map is foreshadowed in the twelve winds around the edge, Noah's Ark, Alexander's camp, the desert monasteries of Egypt and a series of exotic animals along the southern edge of Africa

We cannot suppose that people between the fifth century and the twelfth themselves mapped any substantial part of the known world. There was no reason why they should do so, and in any case the techniques of surveying and of calculating geographical co-ordinates had fallen into disuse, the necessary administrative structure no longer existed and, most important of all, the idea of mapping to scale, of mapping from measurements, had been lost. It follows that where any portion of a world map of this period has a recognisably correct outline, this must go back to a Roman original, and that this original was probably a measured and reasonably accurate map of the world, showing coastal outlines, mountains, rivers, towns and boundaries of provinces.

Serious doubt has recently been cast on the Romans' cartographic awareness – how far they really made and used maps.[73] The only world map we have from this period is the so-called Peutinger Table, compiled probably in the fourth or fifth century but known to us through a copy of about 1200. We would be more likely to call it a diagram than a map; it shows routes, with stopping-places and the distances between them, across the whole of the Roman Empire from Britain to India, but compressed into the form of a strip some 23 feet (7 metres) long and only 14 inches (36 centimetres) wide. We might see it as a series of itineraries in graphic form, accurate and excellent for travelling along the predetermined route from place to place, but of no value if one tried to cross country on one's own initiative from one route to another.[74]

However, it seems inescapable that not later than the fourth century copies were in existence of a more conventional world map – or of more

A section of the Peutinger Table. The strip in the centre is Italy, with Rome and Ostia on the right; Spain is below, and above are parts of central Europe and the Balkans

than one – that was based probably on measured survey, possibly on geographical co-ordinates, and that this map contained, at the very least, all the most accurate features that can be detected in any surviving medieval world map. O. A. W. Dilke suggests that the account of this map's creation in the Hereford map is no more than slightly garbled. Two accounts from the late Roman period tell how in 44 BC Julius Caesar sent out four surveyors to measure the world – the three of the Hereford map (but Nicodemus, not Nichodoxus) plus Didymus who measured the western quarter. It was thirty-two years before they all finished, in the reign of Augustus. Their work will have served as the basis for a world map which – as recorded by the nearly contemporary Pliny – the emperor ordered Marcus Vipsanius Agrippa to set up for public display in a colonnade at Rome. However, Kai Brodersen argues that the evidence for Caesar's survey is too late, and the details too questionable, to be believed, and that what Agrippa produced was not a map but a written text, a geographical description with measurements.[75] It seems that for the Roman map of the world we have no firm evidence but its travestied medieval derivatives. However, since these derivatives can have had no other source, we must see their evidence as irrefutable.

The phrase *mappa mundi,* literally cloth of the world, is unknown to classical Latin. It is first recorded in the early ninth century,[76] and it was what a world map was normally called in the twelfth and thirteenth centuries, whether drawn on cloth or not. In applying the phrase to the Hereford map we are calling it what Richard of Holdingham will have called it himself. However, the meaning of the phrase broadened to include other sorts of map – already in the late thirteenth century we find a chart of the Mediterranean referred to as a *mappa mundi*[77] – and from it English and some other languages derive their modern word for maps of all kinds. The full phrase occurs in English in the fourteenth and fifteenth centuries as *mappemounde,* and it still survives in French and Italian as *mappemonde, mappamondo,* meaning solely a medieval map of the world.

The Early Medieval Forebears

On the Hereford map the triangular Sicily is practically the only outline that has survived the process of transmission from classical Antiquity, but a very few other medieval world maps have more convincing details. One is the Cotton map, which dates probably from the second quarter of the

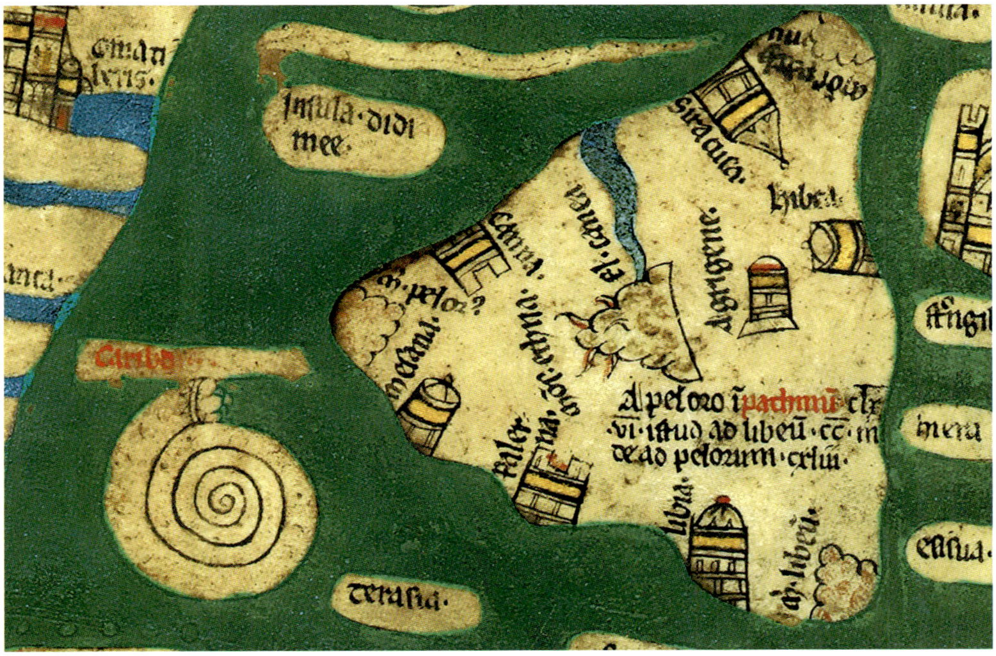

Sicily is to us the Hereford map's only recognisable outline. At its centre Mount Etna spews out flame, and the coil in the sea on the left is the whirlpool Charybdis

The 11th-century Cotton world map, of English origin, preserves a memory of the Roman map of the world in the relative accuracy of the North Sea and English Channel coastlines (bottom left: east is at the top). Boundaries are prominent but few rivers or towns appear

World map in an 11th-century manuscript of Macrobius drawn at Freising. North is at the top, and in the central Mediterranean we see recognisable outlines

eleventh century; it is so called because it is among the manuscripts collected by Sir Robert Cotton (1571–1631) and now in the British Library. Compared with the Hereford map, the relative accuracy of its outlines of Jutland, the Frisian coast, Britain, Ireland, the Peloponnese, the Dardanelles and the Nile delta is striking.[78] Another eleventh-century map with clearly recognisable outlines was drawn at Freising in Bavaria; here we see Corsica and Sardinia in something like their accustomed shapes and we see too the north-west to south-east axis of Italy and the Adriatic.[79] It need not surprise us that these recognisable outlines have only occasionally come down to us. The outlines on medieval world maps are the outcome of many successive copies made by draughtsmen who perhaps had little idea of the nature of cartography and certainly no idea at all of the actual shape of any part of the world. In their hands any error will all too easily have been repeated and magnified – or the outline may even have been deliberately distorted, to force the map into a circle or other arbitrary frame.

There may well have been other reasons for deliberate distortion. In 2002 Brigitte Englisch argued that the outline on five of our earliest world maps, dating from between 775 and 1109, conforms exactly to a predetermined pattern. At its heart is an equilateral triangle; at two of the angles are Constantinople and Carthage, while Jerusalem and Alexandria are in the centre of two of the sides and thus both on a circle drawn within the triangle. These were the four sees of the Church's Patriarchs between 637

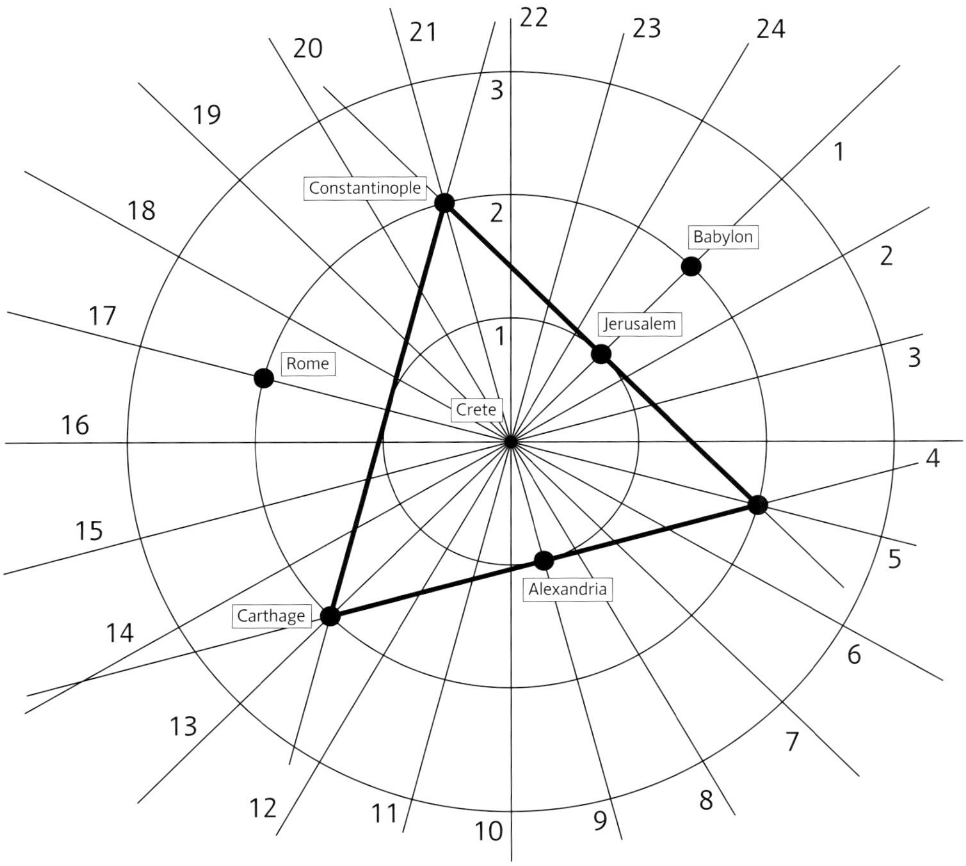

Pattern that Brigitte Englisch sees as the framework for the earliest known medieval maps. North is at the top

and 742. Rome and Babylon both lie on another circle joining the points of the triangle, while a third circle, the same distance beyond the second, marks the edge of the Ocean, the rim of the inhabited world. From the map's centre – in Crete, though this is not significant – twenty-four lines radiate and other cities are placed on them. This pattern was symbolic. The equilateral triangle represented the co-equal Trinity, the three circles divine perfection; the twenty-four lines recalled the number of the heavenly spheres and the twenty-four elders around the throne of God in the book of Revelation, were twice the number of the Apostles and of the Tribes of Israel, and so on.[80]

It is perhaps harder to agree with Brigitte Englisch's further argument that an adapted version of the same pattern underlay many later medieval world maps – indeed, she sees it in the Hereford map, with a right-angled triangle at the centre and no fewer than ten concentric circles, which extend beyond the edge of the map itself to pass through the figures of Mary and Christ outside it.[81] But the details of the pattern she finds, her

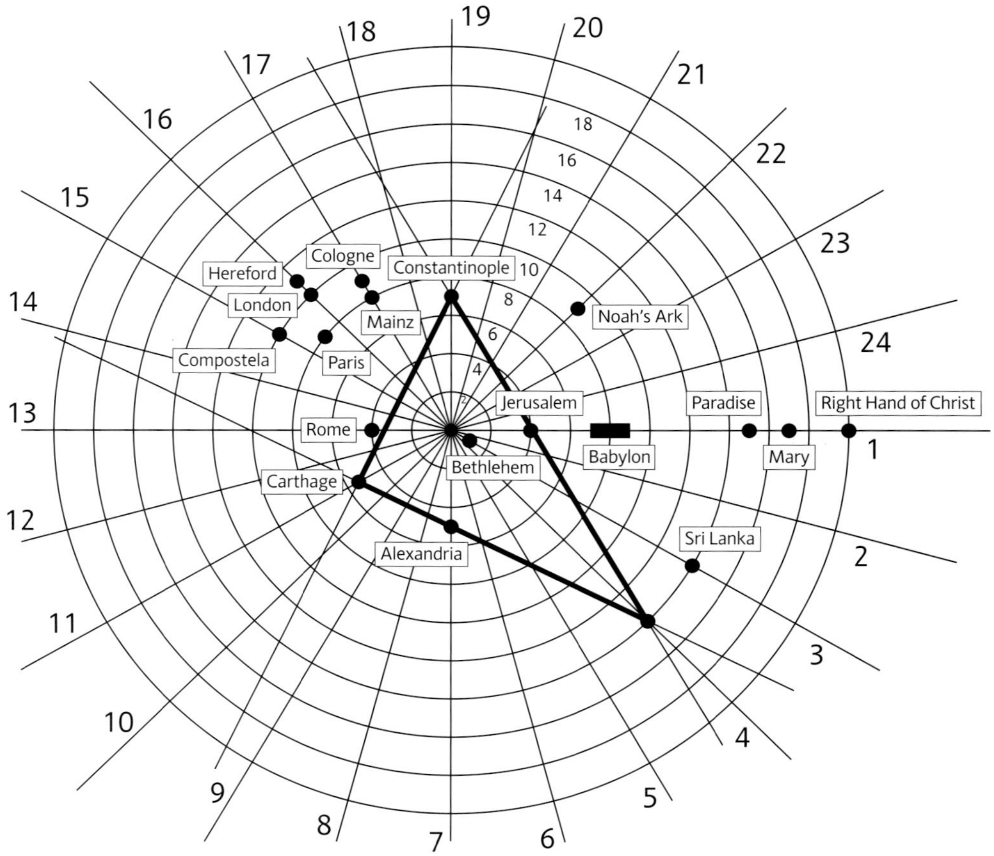

Pattern that Brigitte Englisch sees as the framework for the Hereford map. North is at the top

exact interpretation of the maps' layout, hardly matter. What is important is her wholly convincing argument that it is likely to have been this way of thought that underlay the construction of the earliest medieval world maps. Just as a historian then would select material and present the facts to demonstrate the history of God's salvation, so the geographer would produce the map to display God's carefully regulated ordering of the world. The map had a more elevated purpose than merely copying geographical outlines; it sets before us the harmony of time and space, a statement of divine order in the world, to be sought in its physical form, its geography, as much as in its history, the sequence of events.

If we accept some such process as this, it becomes much easier to see how the probably quite good maps of the Roman period changed so rapidly into the wholly unrecognisable outlines of some of the earliest medieval maps. We would say they degenerated, but people of the time would say they were enhanced, offering a new and vastly more spiritual view of the purpose of a map. Nor need it have been a once-for-all process.

Map of Europe by Lambert of Saint-Omer, early 12th century. The broad peninsula, top right, is Italy with the Alps; bottom right are Spain and Portugal with the Pyrenees

Besides his world map, Lambert of Saint-Omer drew a map of Europe; this takes the form of four thin peninsulas and one broad one, and Hartmut Kugler suggests that it represents the hand of God raised in benediction.[82] In an interesting article Massimo Rossi shows how the sinuous curve of the Mediterranean's north coast on the Hereford map exactly replicates the shape of Christ's body on the Cross at Jerusalem and may very likely have been intended to echo the presence of the crucified Christ in the world.[83] Again, in relation to other places in Palestine Jerusalem is well to the north of where it should be, moved presumably to bring it to the centre of the map.[84] Certainly we miss the whole point of the Hereford map and of other medieval world maps if we see them as offering no more than distorted geographical outlines.

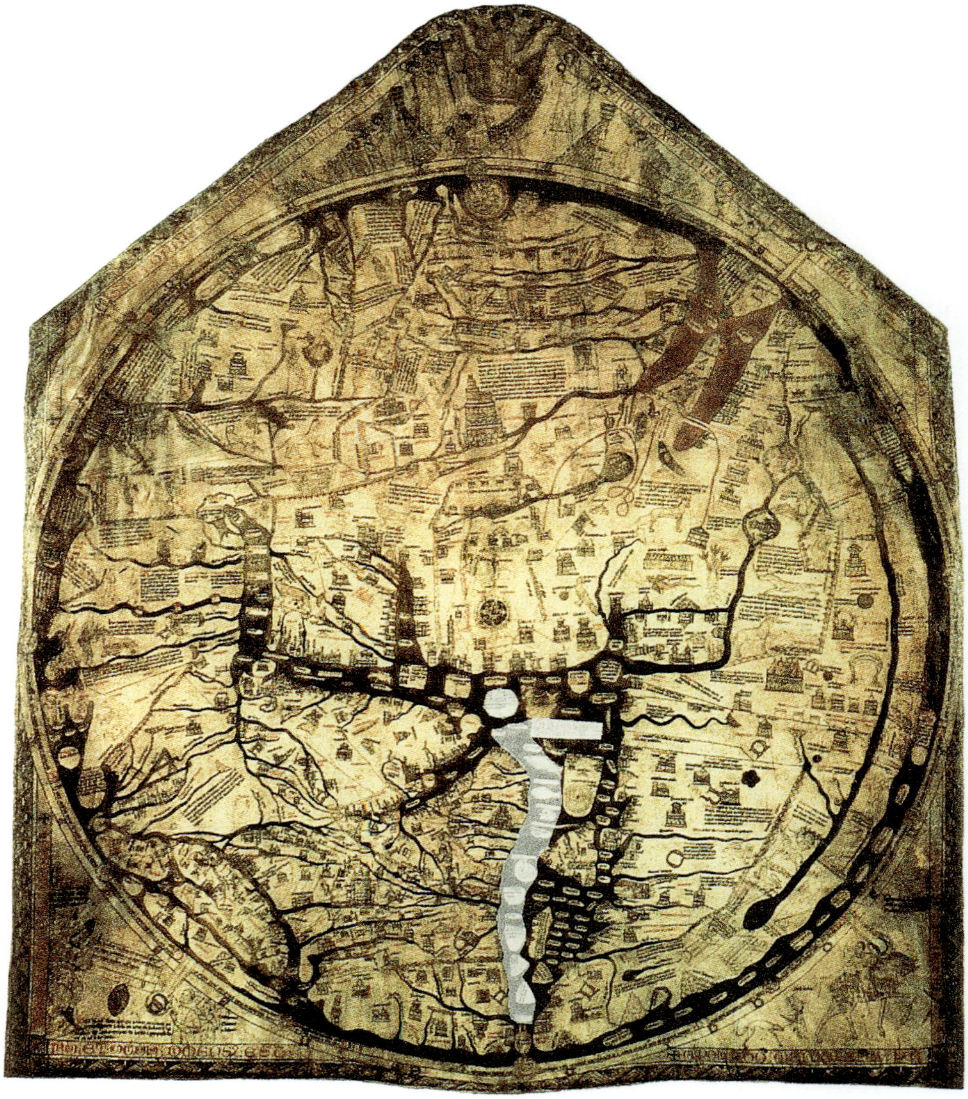

Shape of Christ's body on the Cross, as shown at Jerusalem on the Hereford map, echoed in the map's northern coastline of the Mediterranean

The English Family of World Maps

The early eleventh-century Cotton world map, with its all but recognisable British Isles and other features, is in a manuscript volume that contains, suitably enough, a copy of Priscian's fifth-century Latin translation of the *Periegesis*; this is a description of the world written in Greek verse by an author named Dionysius. Map and text survive together only in this one manuscript, and it may well have been the copyist's idea to unite them. The map is sometimes called the Anglo-Saxon map, and its spelling of the name it gives the Bretons – '*Suðbryttas*', using a runic letter for *th* – shows that it was of English origin.

This is especially interesting. It may well be that in early medieval England there were more copies of the Roman world map, in fully detailed form, than in other areas of Europe. Certainly in the twelfth and thirteenth centuries immensely detailed world maps developed as a genre which, while not exclusively English, seems to have been centred on England and northern France. However, the genre probably had its origins in France. In the early twelfth century a large world map, which included information on the peoples and the natural history of distant lands, was described in detail in a little work by Hugh of Saint-Victor, of the Abbey of Saint-Victor in Paris; it may have been a lecture written down by a pupil.[85] The twelfth-century Munich copy of Isidore of Seville's work, with elaborated map, was produced in northern France, and the so-called Jerome maps of Asia and of Palestine are in a late-twelfth-century manuscript from Tournai, now just across the border in Belgium.[86] In the church at Chalivoy-Milon, 25 miles (40 kilometres) from Bourges, a vast world map, 20 feet (6 metres) across, was painted in the mid-twelfth century. It was destroyed in 1885, and we know of it only through two short descriptions written when it was already much decayed, but from them it sounds very like one of this same group of maps.[87]

However, all other maps in the group were either drawn in England or may just possibly have English associations. We know of some that have failed to survive. There was a *mappa mundi* in Lincoln Cathedral library in the mid-twelfth century and another was among books that Bishop Hugh le Puiset left to Durham Cathedral Priory in 1195. Henry III had world maps painted on walls at Westminster Palace in the 1230s and at Winchester Castle in 1239, and we have references to a map of the world painted on cloth that belonged to Edward I in the 1290s and to another, probably on parchment, that Edward II had acquired by 1324. In the mid-thirteenth century Matthew Paris, a monk at St Albans, mentions one at Waltham Abbey and another by the otherwise unknown Robert of Mentley. Of the late twelfth- or thirteenth-century world maps that survived to be known to us in detail, two are in books and three are separate large maps, like the Hereford map itself.[88]

Earlier than any of the English maps we know in this group, but more closely related to the Hereford map than any of them, is a highly relevant written text. Patrick Gautier Dalché has done much important work on medieval texts that are related to maps, above all in his seminal edition of Hugh of Saint-Victor's work. In 2001 he published a rather later description of a world map, entitled *Expositio Mappe Mundi*, 'Account of the Map of the World'; this is known only from two copies that date from the fifteenth century, but the text was written in the late twelfth century in England, probably in Yorkshire and very likely, he suggests, by the chronicler Roger of Howden. The text as we have it is incomplete; it may never have covered areas it omits in Asia, but references in the text itself show that as originally written it included parts of western and northern Europe that are now missing. However, its relationship to the Hereford map is both clear and close. Scott Westrem finds that half of all the inscriptions on the map (excluding the parts certainly lost from the surviving copies of the *Expositio*) appear in exactly, or very nearly exactly, the same form in the text, while there are others for which the match is less precise.[89]

However, the *Expositio* cannot have been much earlier than the Sawley map; this used to be called the Henry of Mainz map, a name that arose through a misunderstanding. The map accompanies a copy of the early twelfth-century encyclopaedia by Honorius Augustodunensis – sometimes called Honorius of Autun, though it is not certain that this is what his Latin name means. This copy was made probably at Durham in the late twelfth century and was then, with other books, given to the struggling Cistercian abbey at Sawley in Yorkshire. Alternatively, the map may have been drawn at Sawley but, if so, it was almost certainly copied from a Durham original. There is no reason to connect the map with the book's author but every reason to connect it with the map that le Puiset bequeathed to the cathedral priory. Although the map gives little general information and few pictures beyond the symbols for cities, we see links with the Hereford map in, for instance, the island of paradise at the top and the pyramids in Egypt, named on both as Joseph's barns with a drawing of a medieval barn.[90]

More obviously related to the Hereford map is one which is an integral part of a psalter that was produced probably in London in the 1260s. It follows a prefatory series of scenes from the life of Christ that were added to the manuscript a little later, and is itself followed by the calendar which as usual precedes the text of the psalms. It should perhaps be seen as a tiny picture of a large world map. Only 3½ inches (9 centimetres) across, it hardly has space for much information, but it includes fourteen pictures of monstrous races in southern Africa. Like the Hereford map it is a circle centred on Jerusalem with twelve winds around the edge – little human

The world map from Sawley Abbey, Yorkshire, drawn c.1200 probably at Durham, has many details in common with the Hereford map

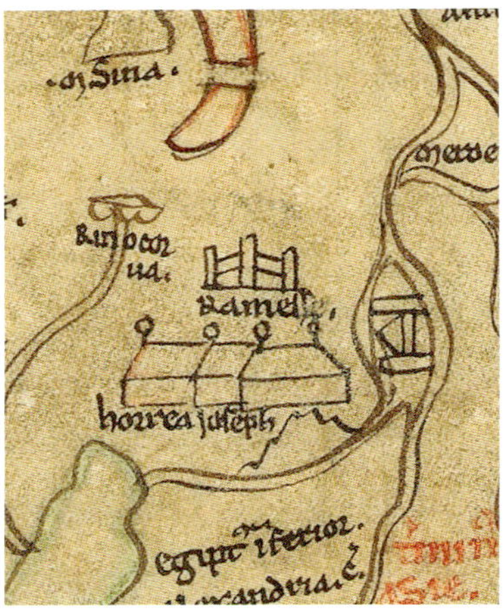

The pyramids of Egypt, identified as the barns in which Joseph stored corn against famine in the book of Genesis, appear as a 13th-century barn on both the Sawley map (left) and the Hereford map (right)

The Psalter world map, of the 1260s, is a tiny picture of a large encyclopaedic map such as the Hereford map with Jerusalem at the centre. East is at the top, and in the earthly paradise are the faces of Adam and Eve

faces blowing hard – and it has a markedly similar outer frame. Above, between two angels swinging censers, God raises the right hand in blessing and in the left hand holds an orb that itself symbolises the world – it is in effect a minute T-O map. Below are two wyverns, two-legged dragons. On the back of the same leaf is a companion piece, a T-O map held before the figure of God with a list of provinces and cities in the space for each continent and again with two wyverns below. The purpose of these two maps is clear from their context: displaying the extent and variety of God's earthly creation they invite reflection on his omnipotence and are didactic in this spiritual sense – that they also convey geographical information was of secondary importance.[91]

The most mysterious map of the group is in the cathedral library at Vercelli in north Italy. On a single piece of parchment and now badly damaged and faded, it was drawn as a near circle, probably slightly flattened at the sides, 33 inches (84 centimetres) from top to bottom. It has been variously assigned an Italian, a Spanish and a French origin, and has been dated either early or late in the thirteenth century as the Philip, king of France, whom it shows in north Africa may be either Philip II (1180–1223) or Philip III (1270–85). In fact it is simplest to connect its presence at Vercelli with the scholarly bishop Gualo Bicchieri, who was papal legate to England in 1216–18 and who brought back with him a collection of manuscripts – among them an outstandingly important volume of Anglo-Saxon

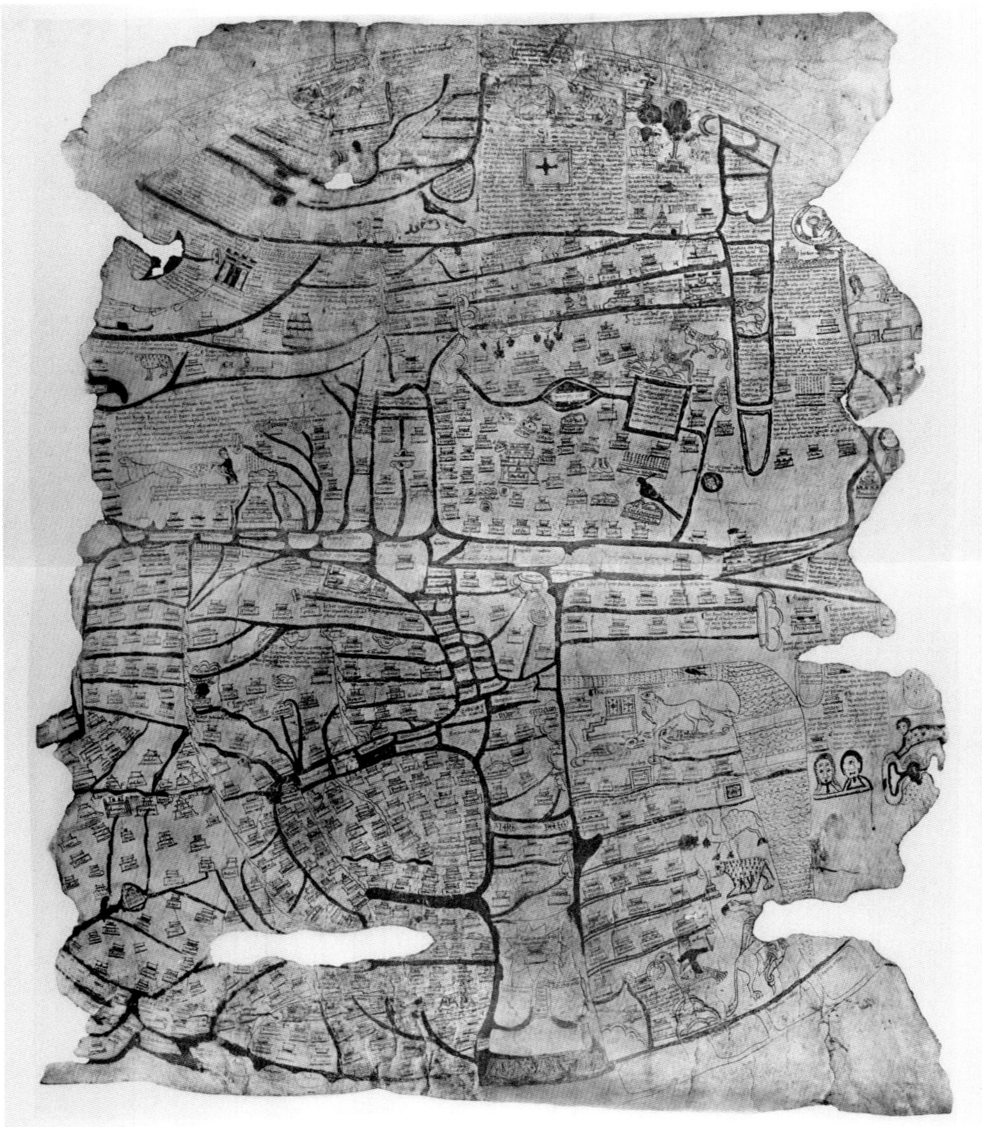

The Vercelli world map, 13th century, may have been meant as a draft for something like the Hereford map. East is at the top, but the Mediterranean islands are drawn excessively large, so that the sea itself is reduced to a series of narrow channels between them

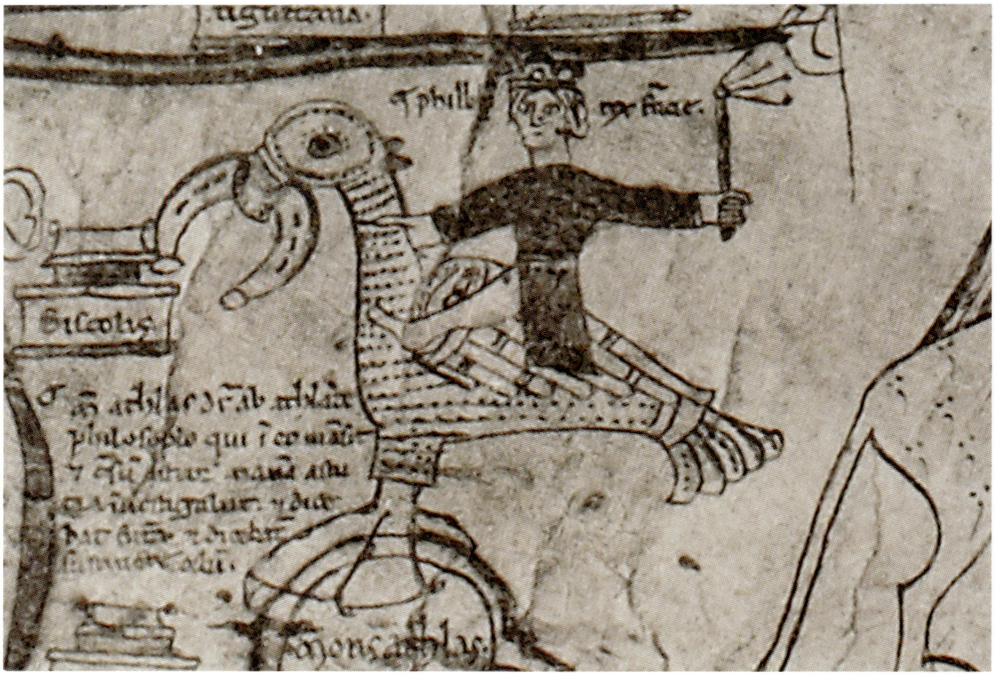

Philip, king of France, pictured and named on the Vercelli world map, sits astride an ostrich; in its beak is a horseshoe that it is about to eat

poetry. C. F. Capello suggests that the map was drawn by the canons of Saint-Victor at Paris, with whom the bishop made contact on his way home, but the picture of King Philip is not necessarily complimentary – he is shown riding an ostrich and wielding a scourge – and the map's apparent omission of the British Isles need not rule out an English origin, for in this part it is badly damaged. However, the map is probably the work of at least two hands, and its origin may well be complex. What is certain is that it is unfinished, and was produced far more roughly than the other world maps in the group. Its outlines were perhaps drawn by someone who did not thoroughly understand the map that was being copied – this could explain how the Mediterranean has been reduced to narrow channels between its many islands. The map may have been brought to Vercelli as a sketch from which a well-executed world map might be made.[92]

The map most often compared with the Hereford map came from the convent at Ebstorf, near Lüneburg in north Germany, and from its local detail there is little doubt that it was made there. It has been dated 1239 from the political implications of what it shows. The prior of Ebstorf from 1223 to 1234 was named Gervase and it has been argued that he was Gervase of Tilbury, of known English origins and connections; even if he was not, the name may still point to English associations, for Gervase was a not uncommon name in contemporary England, while rare in Germany. However, from its style and handwriting the map has been dated to the end of the thirteenth century, and one possibility is that it was a version made at Ebstorf of a map produced earlier elsewhere – just as the Hereford map may be based on an earlier version at Lincoln. The map, since 1835 in the keeping of the Lower Saxony Historical Society (Historischer Verein für Niedersachsen), was deposited in 1906 in the provincial archives

The Ebstorf world map, late 13th century, from the 2007 reconstruction by Hartmut Kugler

Central Europe on the Ebstorf map, from the 2007 reconstruction. Local knowledge appears both at Ebstorf (Ebbekes storp) *itself, lower left, where three small squares mark the five martyrs' graves at the convent, and at Reichenau, further right, where there are three monastic buildings on the island near Konstanz and the source of the Rhine*

at Hanover and there, in 1943, it was destroyed in an air raid, tragically just before final approval was given for its evacuation. In 1953 four replicas on parchment were made from all available evidence, including a coloured lithographic reproduction published in 1896; both the lithograph and the replicas have often been used to illustrate details from the map. In 2007

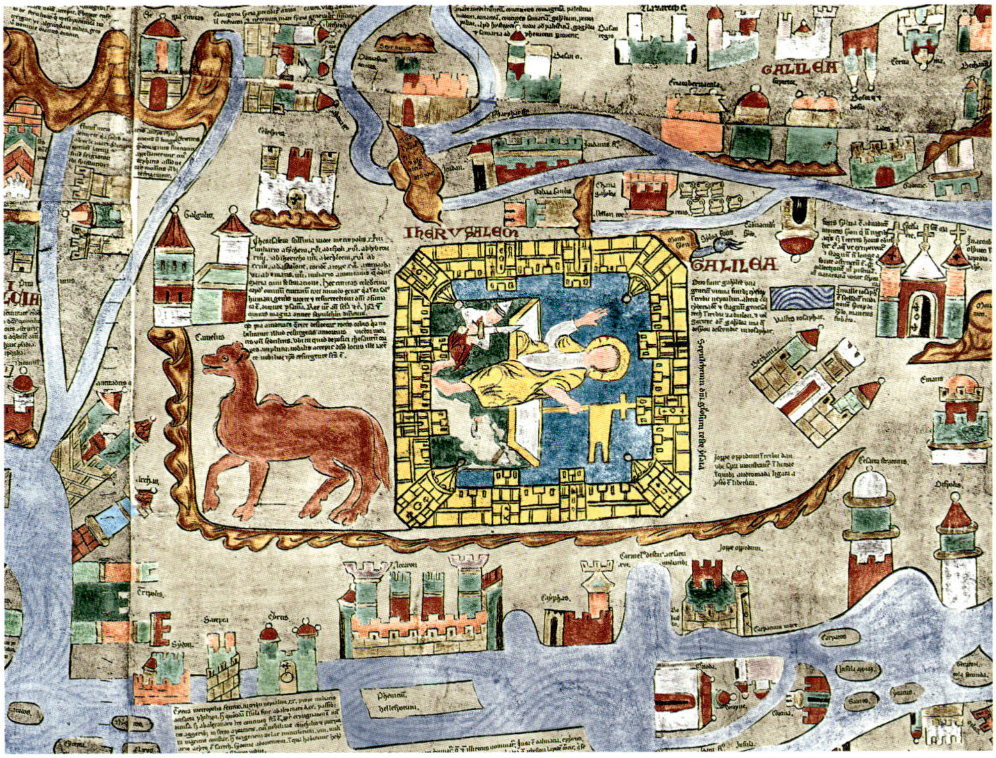

Jerusalem on the Ebstorf world map, from the 2007 reconstruction. Here the city is shown as a walled square, with a picture of the Resurrection, whereas on the Hereford map it has circular walls with the Crucifixion above

Hartmut Kugler, after a new and highly sophisticated assessment of the evidence, published a reproduction of the map that is as accurate a reconstruction as is likely to be achieved; great care was taken over the exact colours and the minutiae of drawing and wording, and with it was published a transcription and discussion of all the wording on the map, comparable to Scott Westrem's work on the Hereford map.[93]

The Ebstorf map was much larger than the Hereford map, some 10 feet (3 metres) square and drawn on thirty pieces of parchment, but it had suffered some damage before it came to light in 1830 – parts of the northern edge of the map were lost and a rectangular portion in the south-east seems to have been cut out deliberately. Intriguingly, it is not impossible that the piece cut out has survived and may someday come to light. The map was circular, with Jerusalem at its centre, and the twelve winds were marked with roundels, left blank, in the surrounding ocean. Also just within the circle of the map were the head, feet and hands of Christ – instead of bearing the world before him, as in the T-O map in the psalter, he is himself fully involved in it. Here again the map displays an aspect of God's relationship with the world. The map was covered with pictures and inscriptions, just like the Hereford map; in the four corners were long extracts from books on geography and natural history and also a note on Julius Caesar's survey of the world, but without naming the surveyors.

The one surviving corner of the Duchy of Cornwall map, the remaining map in the group, is also filled with text, but more closely related to the Hereford map: it names the same three surveyors of the quarters of the world and gives the number of seas, mountains, provinces, cities and so on that were covered by each. All we have of the map is a piece of its bottom right corner, some 2 feet square (62 by 53 centimetres). It is enough, however, to show that the map was circular, slightly larger than the Hereford map and similar in character – we see some of the animals and monstrous races of Africa, pictured and described, and three of what must have been sixteen winds, human heads blowing horns. The fragment survived in the duchy's archives as a wrapper for records of Ashridge College in Hertfordshire; this was a house of Bonhommes, priests living by the Augustinian rule, founded in 1283 under the patronage of Edmund, the king's brother. A book given to the college on its foundation is written by the same hand as the map, which may well have been made and presented at the same time.[94]

In all these world maps we see the map used as a vehicle for varied information arranged geographically. But they are alike not only in concept; they have enough detail in common to show that they are not independent productions – one way or another they are all related. On the other hand they are not simply variant versions of a single map – a glance shows that there are significant differences even in their outline, and in their detailed contents there are far more. To take a few random examples from the two maps of which we know most, the earthly paradise is an island on Hereford, part of the mainland on Ebstorf; Hereford shows the route of the Exodus, Ebstorf does not; in Palestine only a few of the coastal cities are the same on both, while Jerusalem appears as a circle on Hereford, as a square on Ebstorf; the *cynocephali* – the dog-headed people – are placed in Scandinavia on Hereford, in Ethiopia on Ebstorf. These examples could be multiplied many times. That the maps we know are so varied suggests strongly that there were once many others of which we know nothing – we have merely the few survivors. It seems as if, in choosing what details to show, the creator of one of these encyclopaedic maps was likely to draw from a wide pool of material; rather than simply adapting, enhancing or revising an existing map, he would take some information from one map, some from another, some perhaps from a written source that was appearing on a map for the first time. We may be looking at a single family of maps, but the construction of a realistic family tree is scarcely practicable unless possibly on the basis of the geographical outlines rather than their encyclopaedic content.[95]

All this is borne out by contemporary regional maps that are related to the world maps. There are the amoeba-like maps of the British Isles in three copies of the description of Ireland by Gerald of Wales, and a map of Europe in a fourth.[96] There are the maps of Asia and of Palestine in a copy of works on biblical names by St Jerome (*c*.342–420) – they are known as the Jerome maps, but they do not appear in other manuscripts of the same works and they were in fact created at St Martin's Abbey at Tournai in the late twelfth century. The map of Palestine was drawn first with an outline taken from one world map showing the straight Mediterranean coast that is the geographic reality, but was then completely redrawn with deep bights in the coast to north and to south, making Palestine a near peninsula, just as it appears on the Hereford map and on some others.[97] Above all, there are the maps drawn by Matthew Paris, monk of St Albans, writer, artist and man of wide interests. One of his maps is called a world

The so-called Jerome map of Asia, late 12th century, is related to the Hereford map but tells us nothing of distant places and peoples, showing only rivers, mountains, boundaries and towns

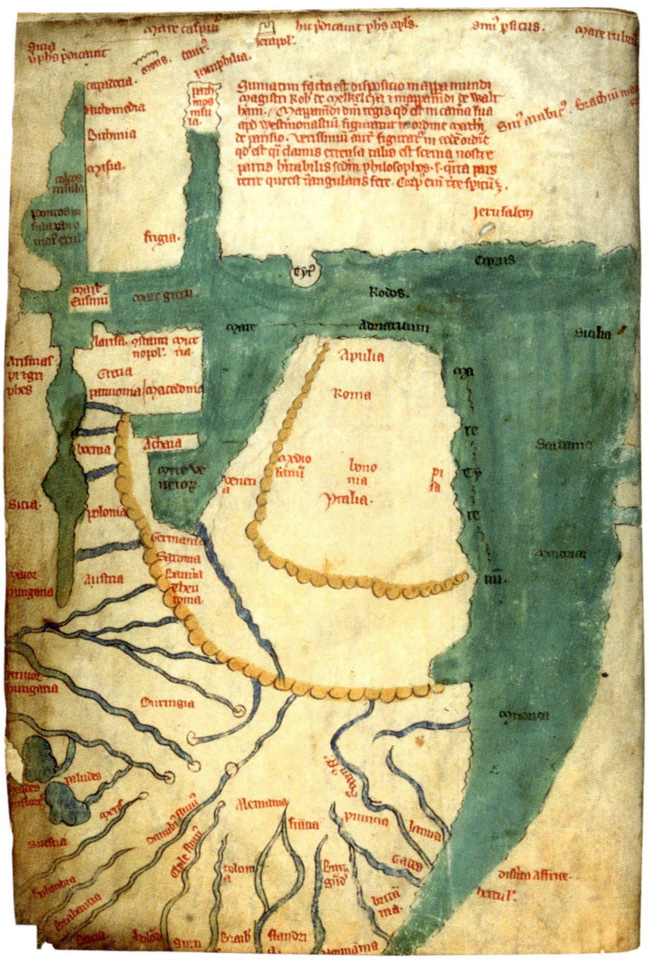

Near-contemporary copy of Matthew Paris's map of Europe and the Near East, mid-13th century. Based probably on a circular world map, it shows Denmark at the bottom on the left, the Straits of Gibraltar on the right, and at the top is Palestine

Two of Matthew Paris's four maps of Britain, mid-13th century. The coastal outlines are taken from world maps on which the shape had not been distorted to fit within a circular frame

map, but it is only part of one, showing Europe and the Mediterranean. The outlines of his four maps of Britain derive from world maps, probably from one with rivers and from another without, and the four maps of Palestine in his hand likewise drew some of their material – coastlines, rivers and place-names – from world maps.[98] The world map may or may not have been a generally familiar object in thirteenth-century England, but it is clear that at the very least an educated person might expect to come across one from time to time.

Nor did the genre die out in the later Middle Ages. Ranulf Higden, who wrote successive versions of his immensely popular *Polychronicon* from the 1330s to the 1360s, included a world map to illustrate its first book on natural history and geography – the other six books are a world history.[99] A closely related map, with England much enlarged, was made as a wall map for Evesham Abbey in about 1390.[100] These maps are entirely in the tradition of the thirteenth-century world maps, and in this they differ from some other late medieval English maps, which draw on the new,

One of Matthew Paris's maps of Palestine, mid-13th century, on which some features derive from world maps. The large walled enclosure is Acre, the last town remaining in crusaders' hands. Jerusalem is top right, a much smaller walled square, and above it are the Dead Sea and the River Jordan

vastly more accurate, portolan charts. These developed as maps for navigation, first in the Mediterranean, then for the more distant places that Italian, Spanish and Portuguese ships visited – at first the Atlantic coast and the English Channel, much later the coast of west Africa. Their influence appears in the Aslake world map. This, like the Duchy of Cornwall map, is no more than a fragment used as a later binding, but it dates from the second half of the fourteenth century, and while it is closely related to the Duchy of Cornwall and Psalter maps it is from the portolan charts that it takes the places it names in north Africa, as well as the Canary Islands beyond.[101] This dual influence appears strikingly in the coastal outline of the mid-fourteenth-century Gough map of Britain, the most remarkable and most mysterious of all medieval English maps in the complexity of its internal detail; the south and south-east coast, the English Channel and the North Sea are of a shape instantly recognisable and taken from the portolan charts, while the western and northern coasts, unknown to Italian navigators, still follow the outline of the older English

World map by Ranulf Higden, monk at Chester, late 14th century. England, bottom left, is given special status, coloured red and covered with cities, but Wales has become a separate island

tradition. By the fifteenth century the most advanced, most innovative maps of the world came not from England but from Italy.

The Hereford map, then, belongs to a whole group of maps which can be seen as a culminating – though sterile – development in a line of tradition

Evesham Abbey world map, c.1390. Jerusalem dominates the centre and at the top the serpent coils around the tree-trunk between Adam and Eve

The Gough map of Britain, mid-14th century, is one of the most remarkable maps from medieval Europe, not least in its network of roads, but where, how and why it was drawn are all uncertain

going back to the Roman period. We probably know of only a tiny proportion of the maps created in this genre in the thirteenth century, when it seems to have been particularly fashionable in England. Of these maps, the Hereford map is now the largest and most detailed survivor – and, like the Ebstorf map, it tells us a great deal about the form and content of those we have never known. It is an invaluable record and relic of an interesting stage of cartographic awareness and development.

4
The Map and Its Sources

The Bible and Early Church History

The Bible contributes less than we might expect to what is shown on the Hereford map.[102] Certainly the events of, in particular, the Gospels were acted out in too small a region to allow much detail to appear. Yet Palestine is disproportionately large on the map and is far from overcrowded – there is more blank space there than in most of Europe or India. It lies, however, in the part of the map – Europe and the lands around the Mediterranean – that shows rivers, mountains and many towns, but not much else. Places from the Old and New Testaments appear in Palestine but we are not shown why they are named, though the picture at Bethlehem incorporates what may be a manger and the well of the oath is marked near the town to which it gave its name, Beersheba. As on the Cotton map, some but not all of the Twelve Tribes are named; on the Sawley map and on some regional maps of Palestine they all appear. But the area is dominated by Jerusalem and by the space created by its surrounds – Mount Ephraim, the Mount of Olives, the Valley of Jehoshaphat – and here we see the Crucifixion, the only event from the Gospels shown on the map. The Ebstorf map shows instead the Resurrection. The walls of Jerusalem were far from circular in either biblical times or the thirteenth century, but we see the same shape on some contemporary plans of the city.[103]

Beyond Palestine the events of Genesis and Exodus are the only part of the Bible to be well represented – we find the same pattern on the Ebstorf map. The terrestrial paradise and the expulsion of Adam and Eve appear in the extreme east of the map, Noah's Ark where it came to rest in the mountains of Armenia. The Tower of Babel dominates the regions between Palestine and India – it dwarfs the city of Babylon beside it – and to its south-east a framed head with battlements above is probably Abraham; the garbled inscription must have been miscopied, but refers to his native city, Ur of the Chaldees.[104] The sites of Sodom and Gomorrah are named in the Dead Sea, and Lot's wife, turned to salt, stands beside – 'very forlorn-looking', say Bevan and Phillott.[105]

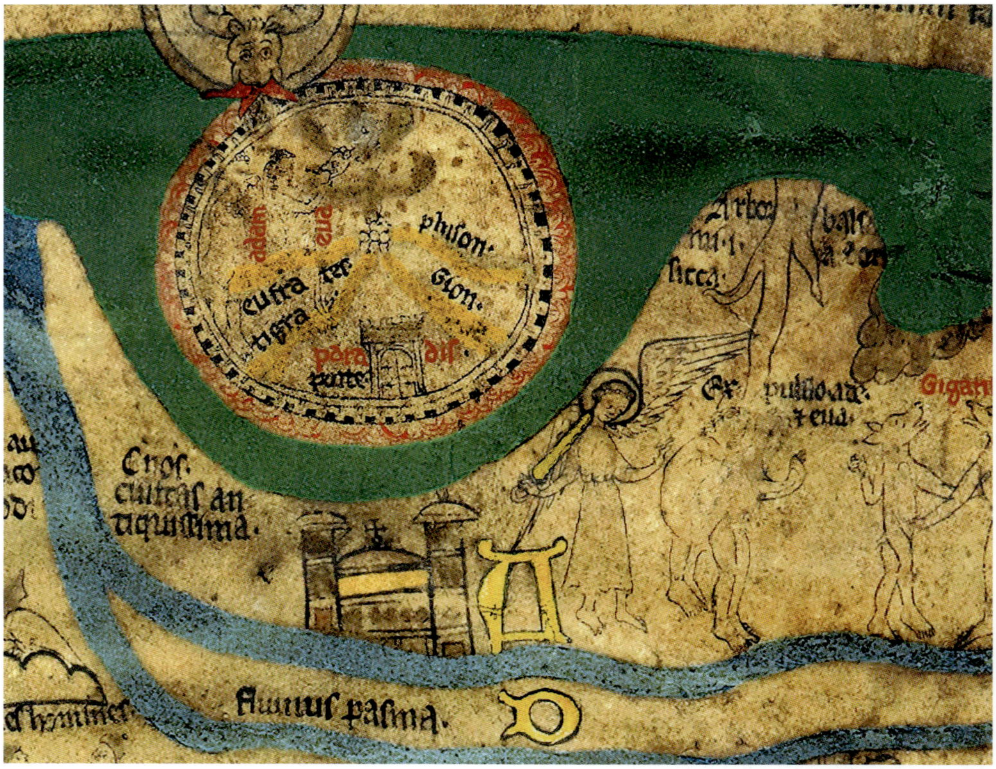

The Garden of Eden is an island with a wall around it at the eastern end of the inhabited world. On the right is the expulsion of Adam and Eve, and above them is the balsam, the tree from one of whose twigs grew the tree used for the Cross

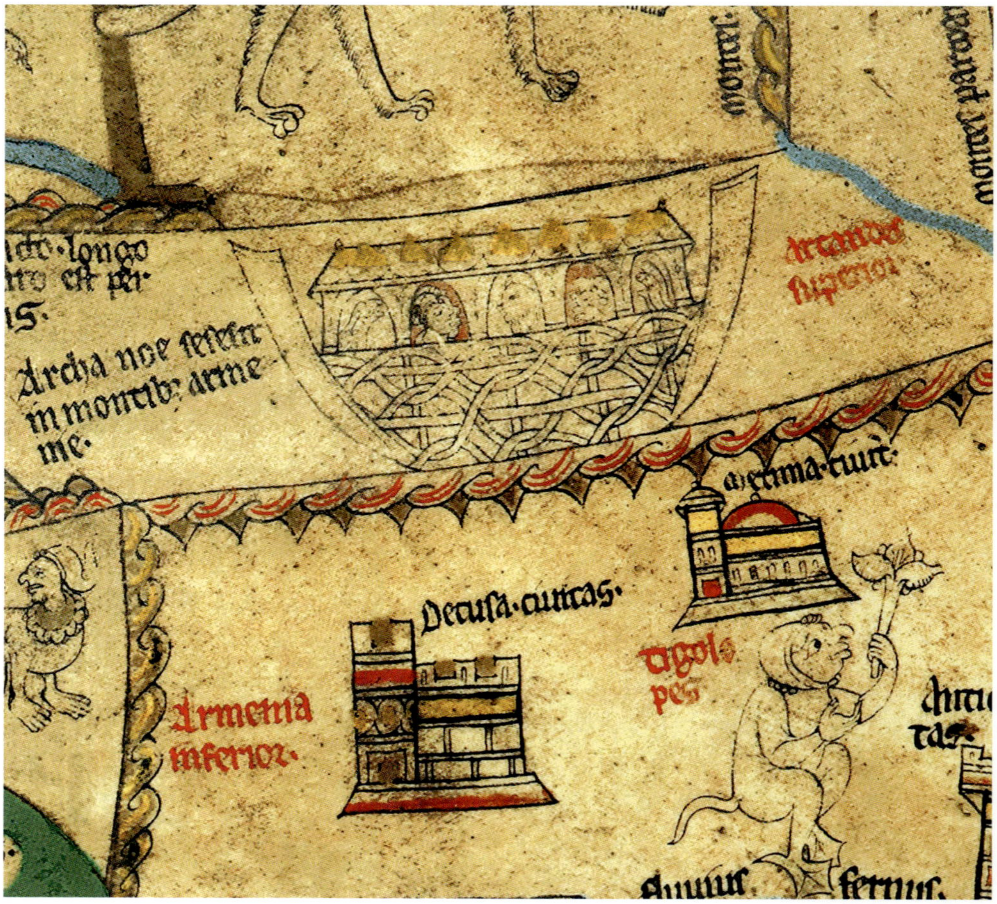

Noah's Ark is shown where it came to rest in the mountains of Armenia. Below is the tigolopes*, a semi-human creature with webbed feet, holding what may be a thyrsus, a staff tipped with a pine-cone ornament and associated with Bacchus*

The Tower of Babel dwarfs the city of Babylon beside it on the right. Further right is a framed head with a garbled inscription that seems to refer to Ur of the Chaldees; this was Abraham's native city and the head is probably his

The book of Exodus is illustrated more comprehensively. In Egypt are Joseph's barns, and the route of the Exodus starts nearby at Rameses. It crosses the Red Sea, and just beyond is Mount Sinai, where Moses is receiving the tablets of the law from the hand of God; the head of Moses has horns, following a common mistranslation of the Vulgate's description of his face as *cornuta,* which really meant radiant. Below are the Israelites worshipping the Golden Calf, shown as a monkey-like creature defecating on the altar and named 'Mahun' (Mohammed). The route then winds and turns to show the wanderings in the desert before it skirts the Dead Sea, crosses the Arnon and the Jordan and ends at Jericho. Other surviving world maps in the English group show places on the path of the Exodus but not the actual line of the route; however, this appears on at least two earlier maps more distantly related[106] – a reminder that the map's closest surviving relatives are no more than a few specimens of what may well have been a large as well as varied family of maps.

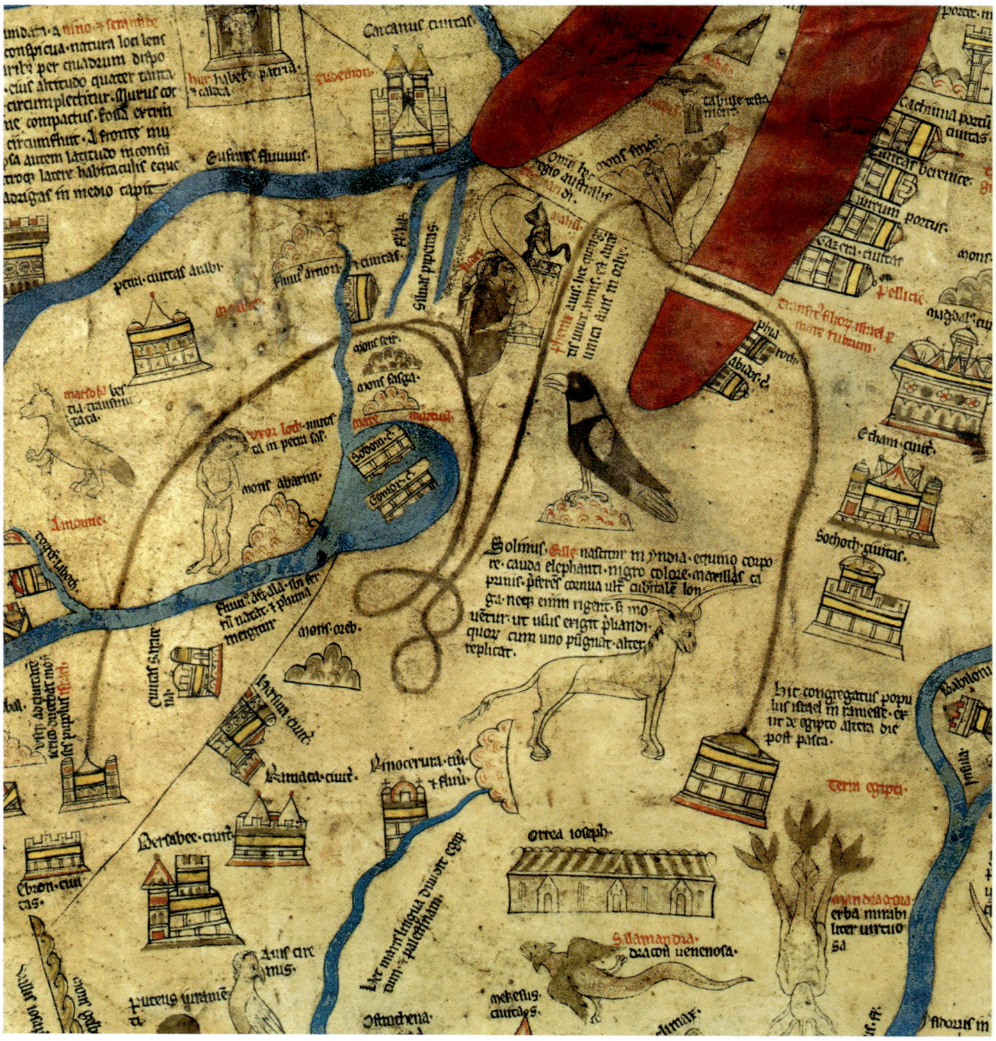

The route of the Exodus crosses the Red Sea to Mount Sinai, where Moses receives the tablets of the law from the hand of God and where (a little further along) the Israelites worship the golden calf. After the loops that mark the wanderings in the desert, the route passes the Dead Sea, with the ruins of Sodom and Gomorrah and Lot's wife, turned to salt, before crossing the River Jordan to reach Jericho

The only other book of the Bible that may have been systematically used for the map is the Acts of the Apostles; towns in Asia Minor and Macedonia, even Reggio and Pozzuoli in Italy, have perhaps been entered from the journeys of St Paul, though as they are dispersed among other places we cannot be sure of this. Of early Christian history we are offered no more than a few fragments. In front of Moses on Mount Sinai is probably the grave of St Catherine of Alexandria, a slab marked with a cross. Far up the Nile two substantial buildings are identified as the monasteries of St Antony (died 356), the desert father venerated as the founder of monasticism, and nearby a large figure in gown and cowl is named Zosimas, a fifth-century monk who retired to the desert. Finally, at Hippo in north Africa St Augustine (354–430) is shown, framed like Abraham at Ur but with spires instead of battlements above. All this amounts to only a tiny part of the information on the map; instruction in biblical and Christian history cannot have been the compiler's chief concern.

St Augustine, doctor of the Church, is shown at his see, Hippo Regius near Bône in Algeria

Geographical Writers

The bulk of the general information on the Hereford map comes from five classical and later Latin authors, writers on geography, history and general knowledge. Before Westrem's definitive work of 2001, much had been done to identify the sources of the map's inscriptions by Bevan and Phillott in 1873; their work was taken further in 1896 by Konrad Miller, who linked them to what appears on other maps. All five authors are in fact identified on the map itself, four of them in the few inscriptions that actually name their source. Thus the description of the rhinoceros comes specifically from Solinus, and the description of the unicorn not only names its author, Isidore, but gives the full reference to his *Etymologies*, book 12, chapter 2. Aethicus and Martianus are similarly named on the map, and outside it, just above the horseman in the bottom right corner, is an inscription naming Orosius: 'Orosius' account of the *Ornesta* of the world, as shown within.' *Ornesta*, correctly *Ormista*, is an acronym used for Orosius's book, *Orosii Mundi Istoria* (History of the World by Orosius). The inscription is seen by Bevan and Phillott as the title to the whole map and by Miller as an acknowledgement of a principal source, parallel to the reference to Augustus' survey in the opposite corner.[107] Neither explanation really fits either its position on the map or its wording – it reads rather like an ill-informed note from the flyleaf of a manuscript containing Orosius' work, and its appearance here is unaccountable.

Orosius was, however, one of the authors most closely connected with the contents of the map. This is not to say that he was himself an original authority. The five writers named were no more than compilers from earlier works – thus Solinus took almost all he wrote from the works of Pliny the elder and Pomponius Mela in the first century AD, two hundred years before him, and they in turn will have used earlier sources. However, Westrem shows that some of the inscriptions on the map are taken directly from Pliny's own work. The *essedones*, who ate their parents' corpses, are described by Solinus and Pomponius Mela – but also by

Herodotus in the fifth century BC. One inscription on the map reveals an authority behind its actual unnamed source, Martianus: in giving the number of towns in the Nile delta it refers back to Artemidorus, a Greek writer of about 100 BC. What these later writers offered were eclectic compilations from a general corpus of information on geography and natural history that originated in the classical period. Nor should we suppose that the map's author himself read these books and constructed the map from what he read – the appearance of so much of the same material on related thirteenth-century maps makes it clear that he did not. He may perhaps have referred back to these sources to revise or add to the information that he found on earlier maps but there was nothing innovative in his reliance on these writers, whether directly or at second hand.

Of Gaius Julius Solinus we know only that he wrote probably in the early third century. He called his book *Collections of Memorable Matters* and from it the map derives much of its information on the monstrous races and strange customs of the most distant parts of the world. Of Paul Orosius we know rather more – he was a pupil of St Augustine at Hippo, having fled

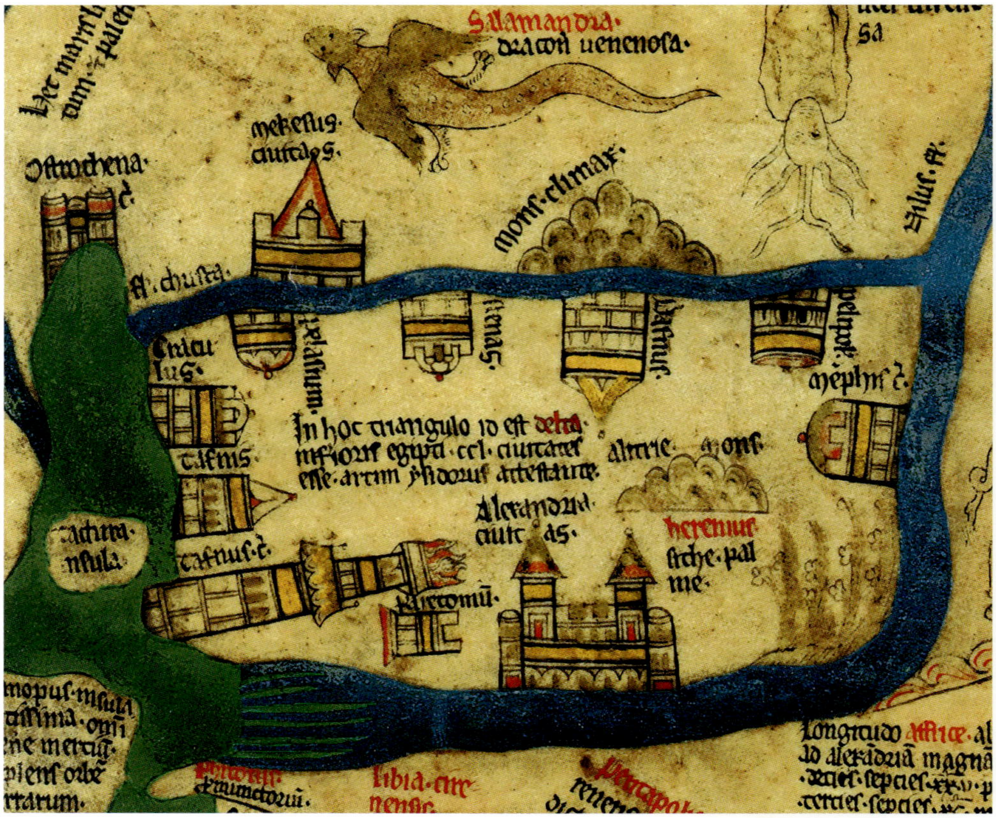

The island between two branches of the Nile is described as part of the delta, where according to Artemidorus there were 250 towns. On the sea by the lower branch is Alexandria, shown with its famous lighthouse

from Spain before the barbarian invasion. His *Histories* were probably completed in 418, and from them the map may have drawn not only its longest inscription – its account of Babylon is copied directly from Orosius – but also much purely geographical information, including rivers and mountains in Asia and Africa. However, Miller argues that this is to see things the wrong way round; he suggests that Orosius himself had access to an ancestor of the map and took his own geographical information from it.[108]

The encyclopaedic work of Martianus Capella, known as the *Marriage of Mercury and Philology*, also dates from the early fifth century, and from this the map takes three of its inscriptions – in the Nile delta, Persia and north-east Asia. From the work ascribed to Aethicus of Istria, which probably dates from the mid-eighth century, the map takes much of its account of the peoples of northern Asia. The map's connection with the work of Isidore, bishop of Seville from about 600 to 636, is clear but imprecise – thus, for instance, the account of each wind on the map can be found in one or other of his two relevant books, the *Etymologies* and *Nature of Things*, but is differently phrased. Indeed, besides identifiable quotations from all these five authors there is much on the map that is clearly related to writings of one or more of them but without following their exact wording. Much of what the map displays is from the general fund of information on geography and natural history that was current in the late classical and early medieval periods.

The eale, *with its horse's body, elephant's tail and goat's jaw, has horns a cubit long, so flexible that they can be used simultaneously in attack and defence. This is one of the few inscriptions that names its source, the 3rd-century author Solinus*

The griste, *in extreme north-east Europe, used their enemies' skins as coverings for their horses and themselves, an account that combines the descriptions of the* griphe *by Aethicus and of the* geloni *by Solinus. The* cynocephali, *the dog-headed people on the left, appear on other maps in India or Ethiopia – the Hereford map, like Aethicus, places them in Scandinavia*

'Here I found beasts like minotaurs, useful for war', a note based on Aethicus that is the map's only use of the first person singular. The creature, part human, part bovine, holds its tail with one hand while making a conversational point with the other. The word massagete*, badly rubbed on the map, has been rewritten*

*At the mouth of the Mediterranean the Pillars of Hercules are shown, literally as pillars, on the island of Cadiz. On the right Gibraltar (*Mons Calpel*) is placed in Africa instead of Europe and beyond is the northernmost of the Fortunate Isles*

Between the upper Nile and the ocean, the map shows a series of ten strange races, among them the people with only one leg and one eye (top), those who cannot open their mouth so have to feed themselves through a straw (second down), those who walk on all fours (fourth down) and two peoples with their face in their chest (sixth and seventh down)

Some other authors identifiably used for the map made more distinctive contributions. The monasteries of St Antony and the nearby faun and satyr that he met in the Egyptian desert derive ultimately from St Jerome's life of St Paul – the fourth-century hermit, not the apostle. The eighth-century *Acts of the Lombards* by Paul the Deacon is the source for the reference on Jutland to the seven sleepers, believed to be Romans from their dress, and also for the two features, probably both shown as heads of dogs, that mark whirlpools at the north and south ends of Britain; they are both named Suilla, hinting at confusion between the Scilly Isles and Scylla and Charybdis in the Straits of Messina – Scylla too is named Suilla on the map. The legend of St Brendan's voyages, as recorded in the tenth or eleventh century, contributed the islands of birds and of rams between Scotland and Ireland and the six Fortunate Isles, south of Gibraltar, which the map calls the isles of St Brendan.

While there has been much learned investigation into the literary and geographical sources of the Hereford and other world maps and into the links between them, there has been no research at all into the pictures that accompany so many of the inscriptions. That here too there are links can hardly be doubted. To take one example, the two-humped camel of the

The camel on the Hereford map (top left), the Ebstorf map (top right), one of Matthew Paris's maps of Palestine (lower left) and the Vercelli map (lower right)

The bird on Mount Gilboa on the Vercelli map (left) and one of Matthew Paris's maps of Palestine (centre), and the phoenix on the Hereford map (right)

Ebstorf map is probably related to the two-humped camel at Hereford – both are walking just east of Acre – and is almost certainly related to the camel on one of Matthew Paris's two maps of Palestine. On the other hand the single-humped camel of the Vercelli map is clearly a different creature altogether. Again, the bird that perches, for no obvious reason, on Mount Gilboa on the Vercelli map and on Matthew Paris's other map of Palestine is very like the phoenix, also perched on a hill, south of the Dead Sea on the Hereford map. We may well wonder if this is not the same bird, used by map-makers in two different roles. The interconnections of these pictures, to say nothing of their sources, could well teach us much about the genesis of this group of world maps and of the relationships between them.

So too might research on the symbols that mark towns on the Hereford and other medieval maps. The elaborate pictures of fantastic buildings (*overleaf*) that are used on the Hereford map are very like those on the Ebstorf map, but with differences in detail, and we can say the same of some fourteenth-century maps of Palestine where we find the same symbol. On the Munich and Vercelli maps towns are also marked by single buildings, but in simpler and rather more uniform style, while the symbols on the Cotton map and some of those on the Sawley map seem to derive from the oblique views of square walled towns that we find in some late Roman works. Many aspects of the graphics of the Hereford map would well repay investigation; they deserve no less attention than has been given to the inscriptions.

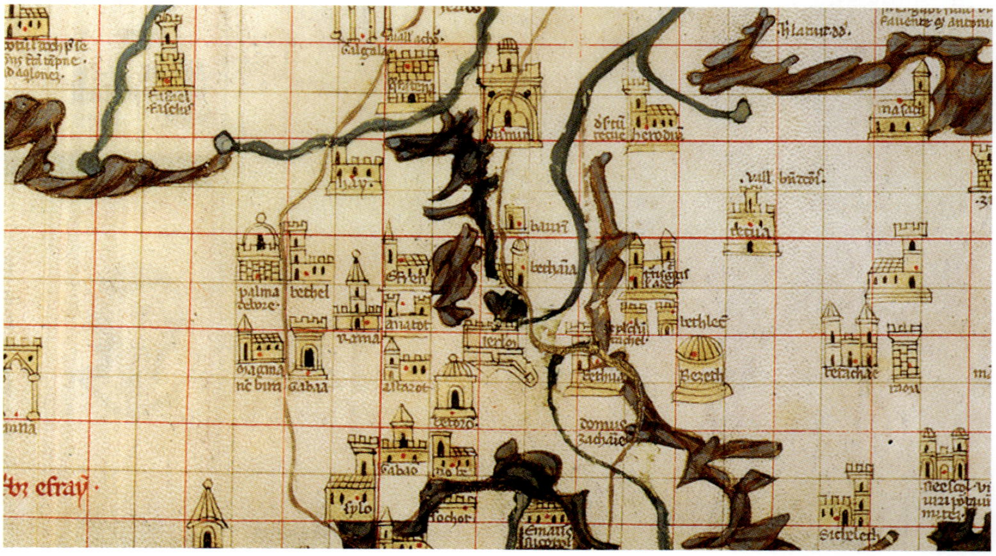

The town symbols on the Hereford map (top), the Ebstorf map (middle) and a map of Palestine of c.1329 (bottom) are similar in character but have never been investigated in detail

Itineraries

The Antonine Itinerary was first compiled probably in the early third century AD; it consists of many detailed lists of places along routes through the whole Roman empire – thus there are some fifteen routes through Britain. Westrem agrees with Bevan and Phillott in naming it among the sources used for the map, linking it with the map's place-names particularly in north Africa. Miller, however, rejected any connection, arguing that the simple posting stations and notes of distances characteristic of the itineraries do not appear on medieval maps; inevitably they show some of the places named in the itineraries, but we should see no direct textual link. In two discussions of the Hereford map in 1954 and 1965 G. R. Crone took the opposite view. He saw the map as a mass of itineraries with many or most of its towns entered from lists of places along routes – the route of the Exodus, the travels of St Paul, routes through the Roman Empire from the Antonine and related itineraries, routes followed by medieval pilgrims and traders.[109]

Behind these two opposed views lie entirely different assumptions about what was shown on the Hereford map's early ancestors. Miller, as we have seen, thought it more likely that Orosius took his information about rivers and mountains from an earlier map than that later maps drew substantially on Orosius as a source. In fact he went further than this, suggesting that much of the general information on the Hereford map and its relatives had already been entered on maps in Antiquity; the thirteenth-century map-makers may well have edited and revised, corrected and added to this material from literary sources, but they started with a great deal of information that was on the earlier maps they copied. Crone on the other hand assumed that with few exceptions even towns were added to world maps only in successive stages from the fourth century to the eleventh; the original maps showed little more than coastal outlines, with provincial boundaries and a couple of cities in each province, and were simply used as a framework to contain an encyclopaedic mass of information drawn from whatever sources were available.[110]

These are both extreme views. A sensible guess – it is no more – is that our map's Roman archetype showed coastal outlines, rivers, mountains, provincial boundaries and also a significant number of towns, but nothing else. There seems no reason why towns should have been left out, the more

The Adriatic. Venice appears as an island nearly opposite Delphi (Delos, *upper right*)*, where a man's face marks the oracle of Apollo. How far towns on medieval world maps derive, like the outlines, from Roman originals is unclear, but some, like Venice, were later foundations*

so as they appear in Roman itineraries, both written and graphic. On this assumption the Cotton map, despite the merits of its coastal outline, would be simply a summary sketch with many omissions, while the Jerome maps would offer a more complete picture, however much distorted, of what appeared on the maps of imperial Rome – we may note that although they contain no encyclopaedic information they still have no awkward blank spaces, even in central Asia. There would be no need for a massive

South-west France. Along the bottom are the Pyrenees with Pau (Palencia)*; top left is Bordeaux* (Burdegala)*. The places along the Dordogne, placed south of the Garonne – Bourg, Fronsac, Libourne – may come from an itinerary used for carrying English wool from Bordeaux to the Mediterranean*

introduction of towns into the map at any later period – many would be there already.

But the Hereford map shows many towns that did not exist in the Roman period – Venice, Compostela, Dublin, Bremen are examples. Whether or not we accept Crone's view of the use of ancient itineraries in the map, his demonstration of the use of twelfth- and thirteenth-century itineraries is convincing and interesting. Lists of successive places along routes were compiled in the Middle Ages as in Antiquity, and Crone finds traces of them on the map, particularly in France. Here the name of a mountain, '*Recordanorum*', recalls the *voie regordane* of earlier Provençal poetry, a route to Provence from the north of France. Other sequences of names in France show commercial routes to the fairs of Champagne and Flanders and from Bordeaux to the Mediterranean, and pilgrims' routes to St James of Compostela. Crone sees a further trade route eastwards from Cologne, presumably through Gandersheim and Halberstadt to Magdeburg on the Elbe, but if it is there it has gone awry for the few towns entered in Germany seem all but arbitrarily placed; the map's line of towns along the River Weser is not in its actual sequence, and some should not be on the river at all. This is not the map's only quirk in Germany. Between the Weser and Elbe in the north and the Danube's tributaries in the south, is an inexplicable rectilinear pattern – there is nothing comparable anywhere

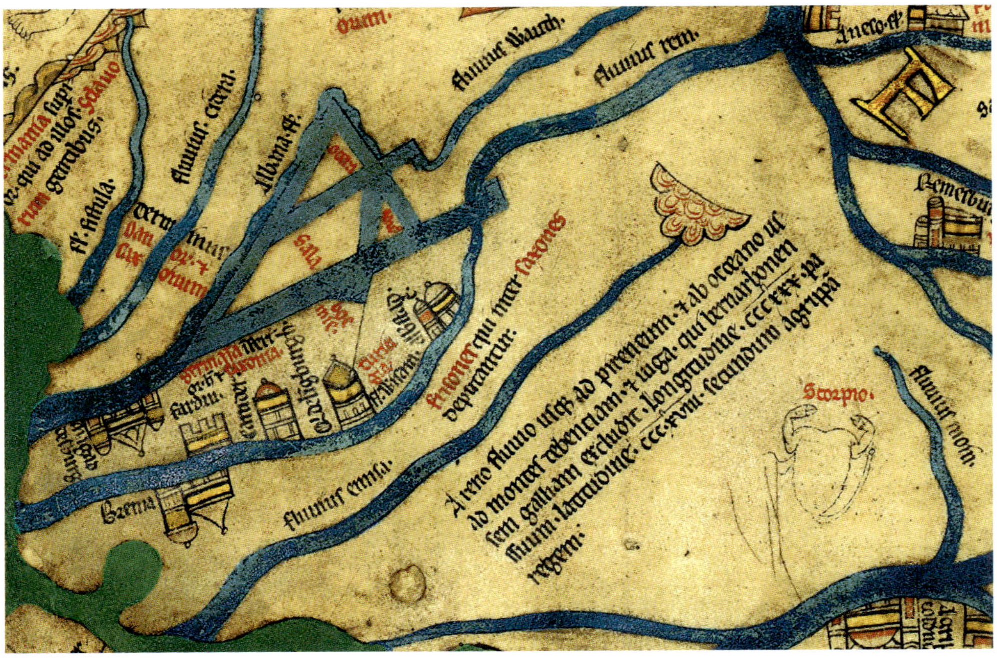

Germany. The Rhine is at the bottom, and the other rivers are successively the Ems, the Weser – with a row of towns – the Elbe, the Eider and the Vistula. The rectilinear pattern between the Weser and the Elbe, connecting with tributaries of the Danube, top right, may represent series of mountains misunderstood by the map-maker

else on the map. The map's draughtsman must have mistaken what he was copying, but what it was – or what he thought it was – is not clear. The lines are not drawn as mountains, though this is possibly what they were on the exemplar. They are not rivers, though the River Elbe hugs their eastern edge, for there is no trace of blue colouring. Miller suggests – but clearly himself without much conviction – that they may be intended as canals linking Elbe and Danube, or as the *limes Saxoniae*, the boundary area on Charlemagne's eastern frontier.[111]

At least sixty-nine of the map's inscriptions can be linked to the supposed journeys of Alexander the Great, as Naomi Reed Kline has shown, and eleven actually mention Alexander: 'Alexander's altars' (*'Aree Alexandri'*), 'The kingdom of Porus and Abisares, who fought against Alexander the Great' (*'Rengnum Phori et Abisaris qui decertaverunt cum magno Alexandro'*) and

The travels of Alexander the Great on the Hereford map, reconstructed by Naomi Reed Kline

so on. They reflect the popularity of the Alexander romances in the thirteenth century; they were names that people might expect to find on the map, with connotations that will have been more obvious then than now. But while the romances undoubtedly influenced the choice of names, they are not themselves the source of any of the inscriptions, which all derive from Solinus, Orosius and other writers. Nor can they have been put on to the map as an itinerary, though from them Naomi Reed Kline has constructed Alexander's travels on the map from Macedonia to Africa, India, the Caspian Gates and Scythia.[112]

Nor is there any sign that the map used itineraries in the British Isles, oddly, perhaps, since Matthew Paris certainly made use of them for his maps of Britain. The river system in England is closely related to what Matthew shows and must derive from the Roman world map, and, as on Matthew's maps, most English cathedral cities are shown. Just possibly Hereford may originally have been one of four left out, along with Chichester, Norwich and Salisbury. As we have seen, the appearance of Caernarvon and Conway reflects current events when the map was being made, and in south-west England Glastonbury and South Cadbury, supposed site of Camelot,[113] are a tribute to fashionable interest in the legends

The British Isles. Though badly distorted to fit within the circular frame, the coastal outline is related to those of the Cotton world map and Matthew Paris's maps of Britain (pp. 45, 62). The towns shown could provide further clues to the connections or interests of the map's author

of King Arthur, just as the Alexander romances left their mark on other parts of the map. Some places, such as Kirkham Abbey in Yorkshire, may have been shown because of some connection with Lincoln or with Richard of Holdingham himself that we know nothing of.[114] In Scotland, as on the River Weser in Germany, the relative positions of the towns, though not quite random, are far from reality. In fact, whatever use he may have made of itineraries for his map, we can be certain that neither Richard nor anyone else envisaged it as a guide to travel. Though a note on the Ebstorf map mentions that this was a possible use for a map, in practice no surviving medieval map earlier than the portolan charts was made to help anyone to get from one place to another.

This was one purpose the map was not meant to serve, but what its intended purpose was, what precisely it was meant to do either at Lincoln or at Hereford we can only guess or infer. What the map does today, however, is to show us how a cultured, well-read person in the England of Edward I might view the world – and, though the colours have faded and much of the visual impact is lost, it does this with a clarity and charm that we can still appreciate seven hundred years later.

Appendix 1

Inscriptions Outside the Frame of the Map

Letters in italics in the transcriptions are expansions of words abbreviated on the map.

1 At the top in the Last Judgement scene

Scroll from the trumpet of the angel on the left
Leuez · si uendrez a ioie pardurable ·
(Rise – you will come to perpetual bliss.)

Scrolls from the hands of Christ
Ecce testimonium meum ·
(Behold my witness.)

Prayer of the Virgin Mary
Veici beu fiz mon piz · de deinz la quele chare preistes ·
E les mamelectes · Dont leit de uirgin queistes ·
Eyez merci de touz · si com uos memes deistes ·
Ke moy ont serui · kant sauueresse me feistes ·
(See, dear son, my bosom within which you became flesh,
and the breasts from which you sought the Virgin's milk.
Have mercy, as you yourself have promised, on all those
who have served me, as you made me the way to salvation.)

Scroll from the trumpet of the angel on the right
Leuez · si alez au fu de enfer estable ·
(Rise – you are going to the fire prepared in hell.)

2 In the bottom left corner

At the top
Lucas in euuangelio · Exiit edictu*m* ab augusto cesare · ut describeretur huniuersus orbis ·

(In the Gospel of Luke: A decree went out from Caesar Augustus that the whole world should be described.)

On the emperor's document

Ite in orbem vniu*er*sum · *et* de omni eius continencia referte ad senatum · *et* ad istam confirmandam Huic scripto sigillum meu*m* apposui ·

(Go into the whole world and report back to Senate on every continent – and to confirm this I have attached my seal to this document.)

On the emperor's seal

S*IGILLVM* · AUGVSTI · CESARIS · IMPARATORIS ·

(Seal of Augustus Caesar, emperor.)

Beside the surveyors

Nichodoxus · Theodocus · Policlitus ·

(Nichodoxus; Theodocus; Policlitus.)

Below the emperor's seal

Tuz ki cest estorie ont · Ou oyront ou lirront ou ueront · Prient a ihesu en deyte · De Richard de haldingham o de Lafford eyt pite · Ki lat fet e compasse · Ki ioie en cel li seit done ·

(All those who possess this work - or who hear, read or see it - pray to Jesus in his godhead to have pity on Richard of Holdingham, or of Sleaford, who made it and set it out, that he may be granted bliss in heaven.)

3 In the bottom right corner

At the top

Descripcio orosii de ornesta mundi · sicut interius ostenditur

(Orosius' account of the *Ornesta* of the world, as shown within.)

Above the huntsman

passe auant ·

(Go ahead.)

INSCRIPTIONS OUTSIDE THE FRAME OF THE MAP

4 On tabs spaced around the map

MORS
(Death)

5 Around the edge of the parchment

In the top left quarter
+ A ⁝ IULIO ⁝ CESARE ⁝ ORBIS ⁝ TERRARVM ⁝ METIRI ⁝ CEPIT ⁝
(Measuring the lands of the earth was undertaken by Julius Caesar.)

In the top right quarter
+ A NICODOXO ⁝ OMNIS ⁝ ORIENS ⁝ DIMENSUS EST ⁝
(All the east was measured by Nichodoxus.)

In the bottom left quarter
+ A TEODOCO ⁝ SEPTEMTRION ⁝ ET ⁝ OCCIDENS ⁝ DIMENS*US* ⁝ EST ⁝
(The north and west were measured by Theodocus.)

In the bottom right quarter
+ A POLICLITO ⁝ MERIDIANA ⁝ PARS ⁝ DIMENSVS ⁝ EST ⁝
(The southern part was measured by Policlitus.)

Appendix 2

Reproductions of the Map

Much of the scholarly work done on the Hereford map has taken as its basis one or other of the copies of the map, published or unpublished. The following are the principal reproductions of the map, with identifying features of those that are printed from drawn copies.

1 (a) London, Royal Geographical Society, Maps, World 448. Manuscript, full size, coloured. Copied 'by Thomas Ballard of Ledbury from the Original in the Chapter House Hereford Cathedral', 1831.

(b) Paris, Bibliothèque nationale de France, Ge A.649. Manuscript, full size, coloured. Copied from 1 (a), 1841.

(c) E.-F. Jomard, *Les Monuments de la géographie ou recueil d'anciennes cartes européennes et orientales* (Paris: Duprat et al., [1842–62]), no. XIV (as eventually issued in the bound volume; nos 1–12 in the original serial publication). Printed, full size, in six sheets, uncoloured lithograph. Copied by E. Rembielinski from 1 (b), lithographed by Kaeppelin. Identifying features: omits the names of 'Carnaruon' and 'Cunweh' in Wales, 'ciuitas bencur' and 'fl. bande' in Ireland; in the inscription referring to Orosius in the bottom right corner 'interius' is spelled 'iterius'.

1 (b) and 1 (c) are faithful copies of 1 (a) and 1 (b), but 1 (a) is far from being a faithful copy of the original. Its most obvious error may be deliberate: the circle of the map itself has been enlarged, so that it protrudes beyond the outer border in the middle of each side and wholly interrupts the border at the bottom. This has consequential changes in the spacing of the inscription around the edge. But besides this, unfamiliar words have been miscopied (e.g. 'essedones' appears as 'Eitedones') and many names are omitted altogether.

2 (a) Photographs, reduced, of the whole map in four sections, uncoloured. By T. Ladmore of Hereford, 1868.

(b) Specimen section of the map circulated with a prospectus for 2 (c), covering part of north-east Asia (the area between the *monoculi* at the top, the pelican bottom left and the inscription about Parthia bottom right). Printed, full size, single sheet, five-colour lithograph. Copied by G. C. Haddon of Hereford but apparently with reference also to 2 (a), lithographed by E. Gailliard & Co. of Bruges, 1869 or 1870.[115] Identifying features: 'ciuitas' is spelled 'ciuicas' after Cristoas, 'ciustas' after Octoricirus.

(c) Published by E. Stanford, London. Printed, full size, six sheets, five-colour lithograph. Edited by F. T. Havergal of Hereford, copied by G. C. Haddon of Hereford, F. Rogers of London and W. Dutton of Hereford but apparently with reference also to 2 (a), lithographed by E. Gailliard & Co. of Bruges, 1872. The area covered by 2 (b) has been redrawn. Identifying features: in the area covered by 2 (b) 'montes' is spelled 'monies' before 'Paropanitates' (the Paro Pamisus Mountains); in Ireland the mountain beside Kildare is shaded with slightly curved, not U-shaped, lines; in the inscription along the edge in the bottom right corner the cross-bar of the T in 'POLICLITO' extends as far to the right as the lower part of the letter.

In what is printed in black and red, 2 (c) is an extremely accurate copy of the map and it is difficult to find variations from the original that serve easily to identify it. The other colours, however, are not only a poor guide to the colours on the original but depart from it in detail, especially in failing to distinguish correctly the colours used for shading in the pictures on the map itself and in the borders.

3 K. Miller, *Mappaemundi: Die ältesten Weltkarten* (6 vols, Stuttgart: J. Roth, 1895–8), vol. 4, inset. Printed, three-sevenths size, single sheet, three-colour lithograph. Edited by K. Miller, lithographed by G. F. Krauss of Stuttgart, 1896. Identifying features: in Scotland and England 'Muneth' and 'bathe' are spelled with *th*, not *þ*; in the inscription referring to Orosius in the bottom right corner 'interius' is spelled 'iterius' but with a line above the *i* standing for the missing *n*.

Though he recognised the inaccuracy of 1 (c), Miller seems to have underestimated the accuracy of 2 (c) while, unfortunately, accepting its colouring as correct. He describes how he used 1 (c) for the outline of 3 – which thus has the same enlarged circle of the map – and arrived at the text of the inscriptions from a comparison of 1 (c), 2 (c) and Bevan and Phillott, entering the readings thus established on 3 as well as in the text of his book.[116] He lists inscriptions which he judged incorrect on the original map and which he has corrected both on 3 and in the text of his book; but he has made further alterations silently on both.[117] The letter-forms in the inscription around the edge of the map differ significantly from those of the original.

4 London, Royal Geographical Society, negative A19922. Photograph, reduced, of the whole map, uncoloured. By C. S. Priestley of Camberwell, *c.*1948.

Taken while the map was in London for conservation work.

5 *Reproductions of Early Maps III: The World Map by Richard of Haldingham in Hereford Cathedral circa A.D. 1285* (London: Royal Geographical Society, 1954). Printed, full size, nine sheets, collotype from photographs.

The first large-size photographic reproduction of the map, and thus the first to be fully reliable. The detail, however, is not as sharply clear as might be wished.

6 (a) British Library, Manuscript Collections, Deposit 8754, Ektachromes 93528–93533. Photographs, 5 × 4 in. (127 × 102 mm) transparencies, of the whole map (93528), of the whole map in four sections and of one detail, coloured. By British Library Photographic Service, 1989–90.

(b) British Library, Manuscript Collections, Deposit 8754, Kodachromes 58575–58588. Photographs, 35 mm transparencies, of the whole map (58575) and of thirteen details, coloured. By British Library Photographic Service, 1989–90.

(c) British Library, Manuscript Collections, Deposit 8754, negatives 93537–93572, 93651. Photographs, 7 × 5 in. (178 × 127 mm), of the whole map (93651), of the whole map in four sections and of thirty-two details, uncoloured. By British Library Photographic Service, 1989–90.

(d) British Library, Manuscript Collections, Deposit 8754, negatives 93573-93601. Photographs, ultra-violet light, 7 × 5 in. (178 × 127 mm), of the whole map in twenty-nine overlapping sections, uncoloured. By British Library Photographic Service, 1989–90.

Although there is variation between different parts of the map and some details do not show up under the ultra-violet light, 6 (d) is sharp and clear. Effectively, however, no more is visible under ultra-violet than under normal light.

7 Dean and Chapter of Hereford. Photograph, transparency, of the whole map, coloured. By G. Taylor, 1990.

Taken on the map's return to Hereford from the British Library.

8 Hereford Mappa Mundi Trustee Co. Ltd. Photographs, transparencies, coloured. By M. Slingsby, 1995.

(a) 10 × 8 in. (254 × 204 mm), of the whole map.

(b) 5 × 4 in. (127 × 102 mm), of the whole map and of the whole map in five sections.

(c) 60 mm, of thirty details.

These are the photographs used for Harvey 1996.

9 Dean and Chapter of Hereford Cathedral. Photographs. By G. Taylor, 2001–2.

(a) 10 × 8 in. (254 × 204 mm) transparencies, four of the whole map, coloured.

(b) 10 × 8 in. (254 × 204 mm) transparencies, thirteen overlapping sections of the map, coloured. (These were used for the cover of Harvey 1996, 2nd edn.)

(c) 60 mm transparencies, thirty-five of the whole map, coloured.

(d) 60 mm transparencies, fifty-two overlapping sections of the map, coloured.

(e) 35 mm negatives, 104 overlapping sections of the map by raking light to show relief, coloured.

(f) 60 mm negatives, twelve of the whole map, uncoloured.

10 The Folio Society. Digital images. By Michael Pollard, 2009.

These are the images used for the present book.

From some of these reproductions derive many others, often reduced in size. Thus Thomas Ladmore, photographer at Hereford, was offering for sale in 1873 photographs of 2 (c) in three sizes: 13 × 11 in. (330 × 279 mm), 11 × 9 in. (279 × 229 mm) and 5½ × 4½ in. (140 × 114 mm).[118] Illustrations of the map in a single publication do not necessarily derive from the same original reproduction and it is often not made clear that illustrations are from a redrawn version and not from the original map. Thus the illustrations in Bevan and Phillott 1873 derive from 2 (c) (frontispiece) and 2 (a) (following p. xlvii); those in Moir and Letts 1955 from 2 (b) (Asia showing camel, pelican, etc.) and 2 (c) (all others); those in Harvey 1991 from 2 (c) (pp. 29, 33) and 6 (a) (p. 31).

The relationship between these reproductions and some derivatives is complicated. Thus, the reduced four-colour facsimile published in 1975 by A.E. Smith (Printers) Ltd of Gloucester takes as its basis a black-and-white photograph of 2 (c), but prints in red the Red Sea and Persian Gulf, bands of decoration in the Last Judgement scene, the name of the Mediterranean and the inscription around the edge of the parchment, all of which have not been taken from 2 (c) nor from a photograph of the original map but have been redrawn.

Select Bibliography

Books Giving a Full Description and Discussion of the Map

BEVAN and PHILLOTT 1873. W. L. Bevan and H. W. Phillott, *Mediaeval Geography: An Essay in Illustration of the Hereford Mappa Mundi* (London: Stanford; Hereford: Jakeman and Jones, 1873; repr. Amsterdam: Meridian, 1969)

CRONE 1954. G. R. Crone, *Memoir* accompanying *Reproductions of Early Maps III: The World Map by Richard of Haldingham in Hereford Cathedral circa A.D. 1285* (London: Royal Geographical Society, 1954)

DESTOMBES 1964. Marcel Destombes (ed.), *Mappemondes A.D. 1200–1500*, Monumenta Cartographica Vetustioris Aevi, 1; *Imago Mundi*, supplement 4 (Amsterdam: Israel, 1964), pp. 197–202

HARVEY 1996. P. D. A. Harvey, *Mappa Mundi: The Hereford World Map* (London: British Library; Toronto: University of Toronto Press, 1996; 2nd edn Hereford: Hereford Cathedral, 2002)

KLINE 2001. Naomi Reed Kline, *Maps of Medieval Thought: The Hereford Paradigm* (Woodbridge: Boydell, 2001)

MILLER 1896. Konrad Miller, *Mappaemundi: Die ältesten Weltkarten*, 6 vols (Stuttgart: Roth, 1895–8), IV (1896)

MORGAN 1988. Nigel Morgan, *Early Gothic Manuscripts II: 1250–1285*, Survey of Manuscripts Illuminated in the British Isles, 4 (London: Harvey Miller, 1988), pp. 195–200

WESTREM 2001. Scott D. Westrem, *The Hereford Map: A Transcription and Translation of the Legends with Commentary*, Terrarum Orbis, 1 (Turnhout: Brepols, 2001)

Brief Guides to the Map

ALINGTON and HARBOUR 1996. Gabriel Alington, illustrations by Dominic Harbour, *The Hereford Mappa Mundi* (Leominster: Fowler Wright, 1996)

JANCEY 1987. Meryl Jancey, *Mappa Mundi: The Map of the World in Hereford Cathedral* (Hereford: Friends of Hereford Cathedral, 1987; new edns 1994, 1995)

MOIR and LETTS 1955. A. L. Moir and Malcolm Letts, *The World Map in Hereford Cathedral; The Pictures in the Hereford Mappa Mundi* (Hereford: Hereford Cathedral, [1955]; 8th edn 1977)

Discussions of the Map's Author and Origins

DENHOLM-YOUNG 1957. N. Denholm-Young, 'The *Mappa Mundi* of Richard of Haldingham at Hereford', *Speculum*, 32 (1957), pp. 307–14; repr. in *Collected Papers of N. Denholm-Young* (Cardiff: University of Wales Press, 1969), pp. 74–82

EMDEN 1957. A. B. Emden, *A Biographical Register of the University of Oxford to A.D.1500*, 3 vols (Oxford: Clarendon Press, 1957–9), I, 556 *sub* De la Bataylle, Richard

FLINT 1998. Valerie I. J. Flint, 'The Hereford Map: Its Author(s), Two Scenes and a Border', *Transactions of the Royal Historical Society*, 6th ser. 8 (1998), pp. 19–44

RAMSAY 1993. Nigel Ramsay, 'Richard of Haldingham', in C. S. Nicholls (ed.), *The Dictionary of National Biography: Missing Persons* (Oxford: Oxford University Press, 1993), pp. 552–3

YATES 1974. W. Nigel Yates, 'The Authorship of the Hereford Mappa Mundi and the Career of Richard de Bello', *Transactions of the Woolhope Naturalists' Field Club*, 41, part 2 (1974), pp. 165–72

Details of the Map's Contents

CRONE 1965. G. R. Crone, 'New Light on the Hereford Map', *Geographical Journal*, 131 (1965), pp. 447–62

GLENN 1986. John Glenn, 'Notes on the *Mappa Mundi* in Hereford Cathedral', in W. M. Ormrod (ed.), *England in the Thirteenth Century: Proceedings of the 1984 Harlaxton Symposium* (Woodbridge: Boydell and Brewer, 1986), pp. 60–3

Discussions of the Map's Setting and Significance

BAILEY 1993. Martin Bailey, 'The *Mappa Mundi* Triptych: The Full Story of the Hereford Cathedral Panels', *Apollo*, 137 (1993), pp. 374–8

KUPFER 1994. Marcia Kupfer, 'Medieval World Maps: Embedded

Images, Interpretive Frames', *Word and Image,* 10 (1994), pp. 262–88
TERKLA 2004. Dan Terkla, 'The Original Placement of the Hereford *Mappa Mundi', Imago Mundi,* 56 (2004), pp. 131–51

Books Setting the Map in the Context of Medieval Cartography in General

BEAZLEY 1897–1906. C. Raymond Beazley, *The Dawn of Modern Geography,* 3 vols (London: Murray, 1897–1906; repr. New York: Smith, 1949), especially II, pp. 465–642

EDSON 1997. Evelyn Edson, *Mapping Time and Space: How Medieval Mapmakers Viewed Their World,* British Library Studies in Map History, 1 (London: British Library, 1997)

HARLEY and WOODWARD 1987. J. B. Harley, D. Woodward and others (eds), *The History of Cartography,* 6 vols (Chicago: University of Chicago Press, 1987–), I: *Cartography in Prehistoric, Ancient, and Medieval Europe and the Mediterranean* (1987), pp. 283–501

HARVEY 1991. P. D. A. Harvey, *Medieval Maps* (London: British Library, 1991)

HARVEY 2006. P. D. A. Harvey (ed.), *The Hereford World Map: Medieval World Maps and Their Context* (London: British Library, 2006)

Notes

[1] The description of the map's colours in Miller 1896, p. 4, is based on the facsimile of 1872. There are in fact no traces of gold on the crown and chain of the Emperor Augustus, nor on the Golden Fleece, which is coloured yellow with black flecks; the route of the Exodus is not white but olive-brown, with black hatching in some places.

[2] And also, of course, in Scott D. Westrem's monumental transcription, translation and discussion of the nearly 1100 inscriptions on the map (see Westrem 2001).

[3] That the words are addressed by the huntsman to the rider (as Jancey 1987, p. 4), not (as Bevan and Phillott 1873, p. 5; Miller 1896, p. 7) by the rider to the huntsman, seems likely from the position and attitude of both figures and from the words' place in the drawing. The words may well have been a well-known call in hunting (as Crone 1965, p. 448), though not specifically mentioned in a near-contemporary treatise (*La Vénerie de Twiti*, ed. Gunnar Tilander, Cynegetica, 2 (Uppsala, 1956)), or possibly a personal motto (as Bevan and Phillott 1873, p. 5; Miller 1896, p. 7; Denholm-Young 1957, p. 312).

[4] Throughout the rest of this chapter I have drawn on my own observation of the map, and I am most grateful to the Mappa Mundi Trustees for making it possible for me to examine it in 1996 by having it removed from its outer case; my investigation was, however, superficial compared with those of Christopher Clarkson, Nigel Morgan and M. B. Parkes in 1999. See n. 5, below.

[5] Their reports are published in Harvey 2006 as Christopher Clarkson, 'The Hereford Map: The First Annual Condition Report' (pp. 95–106), M. B. Parkes, 'The Hereford Map: The Handwriting and Copying of the Text' (pp. 107–17) and Nigel Morgan, 'The Hereford Map: Art-Historical Aspects' (pp. 119–35).

[6] The Hereford map is far from alone in having its dorse inaccessible to scholarly examination. Other examples are three large medieval maps of the Holy Land, framed and glazed: Florence, Biblioteca Medicea Laurenziana, Ashburnham Libri MS 1882; New York, Pierpont Morgan Library, M.877; Oxford, Bodleian Library, MS Douce 389.

7 Bevan and Phillott 1873, pp. 2–3, Miller 1896, p. 6, and Crone 1954, p. 3, take the name to be Richard of Holdingham 'and' of Sleaford, reading the conjunction as *e* (and). It is in fact clearly *o* (or) on the map, but the photograph in Bevan and Phillott 1873, following p. xlvii, shows the conjunction as *e*. The photograph may have been retouched, but as this is also how Sir Frederic Madden, Keeper of Manuscripts at the British Museum, read it when he examined the map in 1855 it seems most likely that simple grime, since cleaned, produced this effect (Martin Bailey, 'The Rediscovery of the Hereford Mappamundi', in Harvey 2006, p. 68).

8 Hermann Kern, *Through the Labyrinth: Design and Meanings over 5,000 Years* (Munich: Prestel, 2000), pp. 144–6, 166–9.

9 By D. E. Greenway and V. I. J. Flint (Flint 1998, p. 26).

10 Yates 1974, pp. 165–72.

11 Yates 1974, p. 172 n. 30.

12 *Times Literary Supplement*, 12 July 1996, p. 14.

13 Parkes, 'Handwriting', pp. 107–9; Morgan, 'Art-Historical Aspects', pp. 122–7, 130–2.

14 Parkes, 'Handwriting', pp. 111–12.

15 In support of this suggestion it is worth noting that in writing the inscriptions the scribe extended words that had been abbreviated on the exemplar (see n. 20, below), and that it would help to explain how the names of Europe and Africa came to be reversed.

16 Clarkson, 'First Condition Report', pp. 104–5.

17 As we shall see (below, p. 81), this artist probably misunderstood his exemplar when he drew a rectilinear pattern in Germany between the Danube tributaries and the Elbe and Weser.

18 Morgan, 'Art-Historical Aspects', p. 121. The same style of drawing cities continued on maps throughout the fourteenth century, e.g. Paris, Bibliothèque nationale de France, MS latin 4939, ff. 10v–11r (*c*.1329), Oxford, Bodleian Library, MS Douce 389 (*c*.1400).

19 M. B. Parkes has found the work of the same scribe in several short medical texts and a table of lunar cycles, dominical letters and the day of Easter that are in Oxford, Bodleian Library, MS Ashmole 399; at least the table was written in about 1299 (Parkes, 'Handwriting', pp. 107–9).

20 We see failures to understand the exemplar in the garbled inscription at Ur, where the errors seem to include writing '*habet*' for '*habre*' (i.e. Abraham),

probably simply an incorrect expansion of '*hab*' on the draft; the reference in the Nile delta to the unfamiliar authority Artemidorus, which has been changed to the more familiar Isidore, but meaninglessly, as '*artim ysidorus*'; and, as originally written, the failure to expand '*anim*' to '*animal*' in the description of the *bonnacon*. The note about Orosius above the horseman in the bottom right corner may also have been entered on the map by mistake, through misunderstanding the draft or other instructions (below, p. 71).

21 Others are the insertion of the *al* in '*animal*' in the description of the *bonnacon* and of the *r* in 'Samarcan' (Samarcand).

22 Flint 1998, p. 23.

23 Work on the map's later history is much indebted to Martin Bailey, 'The Rediscovery of the Hereford Mappamundi', in Harvey 2006, pp. 45–78, which sets out all known references to the map before the publication of Bevan and Phillott 1873.

24 *History from Marble, Compiled in the Reign of Charles II by Thomas Dingley, Gent.*, ed. John Gough Nichols, 2 vols, Camden Society, old series 94, 97 (London: the Society, 1867–8), I, 35–6, 78, clx (Bailey, 'Rediscovery', p. 49).

25 R. Gough, *British Topography*, 2 vols (London: Payne, 1780), I, 71–6 (Bailey, 'Rediscovery', p. 51).

26 Bailey, 'Rediscovery', p. 50.

27 Martin Bailey, 'The Discovery of the Lost Mappamundi Panel', in Harvey 2006, p. 80; Christopher Clarkson, 'The Hereford Map: The First Annual Condition Report', in Harvey 2006, p. 104. The two samples from the planks gave radiocarbon date-ranges of 1270–1400 and 890–1020 (which are not necessarily inconsistent: old wood might have been reused) and the one from the crockets along the top a date-range of 1040–1280.

28 Bailey, 'Panel', p. 83.

29 Bailey, 'Panel', p. 84.

30 This can be clearly seen in the photograph, made in 1868, of the lower left corner of the map in Bevan and Phillott 1873, following p. xlvii, and on a much smaller woodcut of the whole map in *A Handbook to Hereford Cathedral* (London: Murray, 1864), plate XI.

31 Clarkson, 'First Condition Report', p. 105 n. 7. If it is accepted that the parchment was attached to the boards before the map was drawn and

that the wood came from the Hereford region it follows that the drawings as well as the writing on the map were executed at Hereford, not Lincoln, and that the fold in the map was not made when it was transferred from artist to scribe.

32 Bailey, 'Panel', p. 86.

33 Bailey, 'Panel', p. 82.

34 Bevan and Phillott 1873, p. 11; Bailey, 'Panel', p. 87.

35 Bailey, 'Panel', pp. 79–80.

36 As described by Patrick Gautier Dalché, 'Pour une histoire des rapports entre contemplation et cartographie au Moyen Âge', in *Les Méditations cosmographiques à la Renaissance,* Cahiers V. L. Saulnier, 26 (Paris: Presses de l'Université Paris-Sorbonne, 2009), pp. 32–5.

37 Terkla 2004, p. 140.

38 As set out in detail by Bailey, 'Panel', p. 88.

39 Kupfer 1994, pp. 272–6.

40 Flint 1998.

41 M. B. Parkes, 'The Hereford Map: The Handwriting and Copying of the Text', in Harvey 2006, p. 123; Nigel Morgan, 'The Hereford Map: Art-Historical Aspects', in Harvey 2006, p. 115.

42 Morgan, 'Art-Historical Aspects', pp. 133–4 n. 12.

43 Terkla 2004.

44 I am most grateful to Dr de Wesselow for letting me read and cite his unpublished paper.

45 Bevan and Phillott 1873, p. 11. It might, as Thomas de Wesselow suggests, have been the work of a local member or sympathiser of the Templars Order when the French king Philip the Fair arrested all Knights Templar in France in 1307. Another possibility raised by Westrem and taken up by Terkla, is that it was caused by the pointer used by a local guide to explain and expound the map to pilgrims (Westrem 2001, p. xix n. 11; Dan Terkla, 'Informal catachesis and the Hereford mappa mundi', in Robert Bork and Andrea Kann (eds), *The Art, Science and Technology of Medieval Travel,* AVISTA Studies in the History of Medieval Technology, Science and Art, 6 (Aldershot: Ashgate, 2008), p. 134).

46 Clarkson, 'First Condition Report', p. 101; Bailey, 'Rediscovery', p. 53.

47 Clarkson, 'First Condition Report', pp. 101–2.

48 Westrem 2001, pp. 134–5.

NOTES

[49] Bailey, 'Panel', p. 83; Clarkson, 'First Condition Report', p. 102.

[50] [Richard Rawlinson], *The History and Antiquities of the City and Cathedral-Church of Hereford* (London: Gosling, 1717), pp. 64–5; J. Duncomb, *Collections towards the History and Antiquities of the County of Hereford*, 2 vols (Hereford: Allen, 1804–12), I, 587. Harvey 1996, p. 14, recorded the suggestion that the ringed dapplings on the horse were in the same ink as the signature and thus made at the same time; however, Nigel Morgan has found that they are in fact in the same colour as some of the original underdrawing of the map (Morgan, 'Art-Historical Aspects', p. 119).

[51] Bailey, 'Rediscovery', p. 76.

[52] Gough, *British Topography*, I, 71–6, plate V (Bailey, 'Rediscovery', p. 51).

[53] London: Carter, 1780 (Bailey, 'Rediscovery', pp. 50–1). It did not, however, appear on the title page of Carter's *The Ancient Architecture of England*, 2 vols (London: Carter, 1795), as Miller 1896, p. 4 n., discovered; the mistake originated in Bevan and Phillott 1873, p. 11.

[54] Bailey, 'Rediscovery', pp. 52–4.

[55] Bailey, 'Rediscovery', pp. 54–7.

[56] Paris: Beau, 1842 (Bailey, 'Rediscovery', pp. 60–1).

[57] Bailey, 'Rediscovery', pp. 66–72; Bevan and Phillott 1873, p. 11.

[58] Bailey, 'Panel', pp. 86–7.

[59] The progress of the project can be followed in reports in the *Transactions of the Woolhope Naturalists' Field Club*, in the volumes for 1867, pp. 108–10; 1868, p. 238; 1869, pp. 152–4; 1870, pp. 252–3; 1871–3, p. 134 (Bailey, 'Rediscovery', pp. 74–7).

[60] Bailey, 'Rediscovery', p. 47.

[61] Bevan and Phillott 1873; Miller 1896; Westrem 2001.

[62] Bailey, 'Rediscovery', p. 72; Bailey, 'Panel', pp. 87, 92–3 n. 49. Harvey 1996 repeats the understanding that it was exhibited in London at the future Victoria and Albert Museum in 1862, but this was probably mistaken (Bailey, 'Rediscovery', p. 73).

[63] Bailey, 'Panel', p. 87; an account of its cleaning and repair is in *Hereford Cathedral News*, 4 (Christmas 1948), pp. 11–12.

[64] G. R. Crone, *The Hereford World Map* (London: Royal Geographical Society, 1948); Crone 1954.

[65] Letters from Cockerell to Canon Acheson, 22 April, and from Penelope E. Morgan to Cockerell, 1 June. I am most grateful to Professor

Michelle Brown for giving me access to Cockerell's correspondence relating to the map.

66 Jonathan Alexander and Paul Binski (eds), *Age of Chivalry: Art in Plantagenet England 1200–1400* (London: Royal Academy of Arts, 1987), p. 215.

67 Bailey, 'Panel', p. 93 n. 51. *National Heritage Memorial Fund: Eleventh Annual Report 1990–1991* (London: 1991), p. 1.

68 Kline 2001; Westrem 2001.

69 Edson 1997 gives a full account of medieval world maps before (and some after) the Hereford map; there is a far briefer introduction in Harvey 1991. Peter Barber, 'Medieval Maps of the World', in Harvey 2006, pp. 1–43, describes fourteen of the more important, with an illustration and select bibliography for each.

70 Rudolf Simek, 'The Shape of the Earth in the Middle Ages and Medieval Mappaemundi', in Harvey 2006, pp. 293–303.

71 Edson 1997, pp. 149–59; Barber, 'Medieval Maps', pp. 1–4; Alfred Hiatt, *Terra Incognita: Mapping the Antipodes before 1600* (London: British Library, 2008), pp. 84–9.

72 Edson 1997, pp. 105–11, 161–3; Barber, 'Medieval Maps', pp. 8–10.

73 Most comprehensively by K. Brodersen, *Terra Cognita: Studien zur römischen Raumerfassung*, Spudasmata, 59 (Hildesheim: Olm, 1995).

74 O. A. W. Dilke, 'Itineraries and Geographical Maps in the Early and Late Roman Empires', in Harley and Woodward 1987, pp. 238–42.

75 O. A. W. Dilke, 'Maps in the Service of the State: Roman Cartography to the End of the Augustan Era', in Harley and Woodward 1987, pp. 205–9; Brodersen, *Terra Cognita*, pp. 261–87.

76 P. Gautier Dalché, *La 'Descriptio Mappe Mundi' de Hugues de Saint-Victor* (Paris: Études augustiniennes, 1988), pp. 92–3.

77 Guillaume de Nangis, 'Gesta Sanctae Memoriae Ludovici', in Martin Bouquet and others (eds), *Recueil des historiens des Gaules et de la France*, 24 vols (Paris: 1738–1904), xx, p. 444.

78 Edson 1997, pp. 74–80; Barber, 'Medieval Maps', pp. 4–8.

79 Munich, Bayerische Staatsbibliothek, Clm 6362, f. 74r.

80 Brigitte Englisch, *Ordo Orbis Terrae: Die Weltsicht in den Mappae mundi des frühen und hohen Mittelalters*, Vorstellungswelten des Mittelalters, 3 (Berlin: Akademie Verlag, 2002), pp. 123–69.

81 Englisch, *Ordo Orbis Terrae*, pp. 450–67, 648–51.

82 Hartmut Kugler, 'Europa Pars Quarta: Der Teil und das Ganze im

'Liber Floridus', in Ingrid Baumgärtner and Hartmut Kugler (eds), *Europa im Bild des Mittelalters: Kartographische Konzepte*, Vorstellungswelten des Mittelalters, 10 (Berlin: Akademie Verlag, 2008), pp. 55–9.

83 Massimo Rossi, 'The Hereford Mappamundi: Visibile parlare', in Harvey 2006, pp. 185–9.

84 On the Ebstorf map (below, p. 56), which also has Jerusalem at its centre, the displacement is still more marked. This throws an interesting light on the original composition of both maps.

85 Gautier Dalché, *La 'Descriptio Mappe Mundi'*.

86 Edson 1997, pp. 26–30.

87 M. Kupfer, 'The Lost *Mappamundi* at Chalivoy-Milon', *Speculum*, 66 (1991), 540–71.

88 P. D. A. Harvey, 'The Sawley Map and Other World Maps in Twelfth-Century England', *Imago Mundi*, 49 (1997), 36–40; P. D. A. Harvey, 'Maps of the World in the Medieval Royal Wardrobe', in Paul Brand and Sean Cunningham (eds), *Foundations of Medieval Scholarship: Records Edited in Honour of David Crook*, Borthwick Texts and Studies, 36 (York: Borthwick, 2008), pp. 51–5. For the identification of *Melkeleia* with Mentley, in Standon, Hertfordshire, see J. E. B. Gover, Allen Mawer and F. M. Stenton, *The Place-Names of Hertfordshire*, English Place-Name Society, 15 (Cambridge: Cambridge University Press, 1938), pp. 197–8.

89 Patrick Gautier Dalché, *Du Yorkshire à l'Inde: Une 'géographie' urbaine et maritime de la fin du XIIe siècle (Roger de Howden?)*, Hautes études médiévales et modernes, 89 (Geneva: Droz, 2005), especially pp. 49–82; Westrem 2001, pp. xxxiv–xxxvii.

90 Edson 1997, pp. 111–17; Harvey, 'Sawley Map', pp. 33–7; Barber, 'Medieval Maps', pp. 10–13.

91 Edson 1997, pp. 135–7; Barber, 'Medieval Maps', pp. 15–19.

92 C. F. Capello, *Il mappamondo medioevale di Vercelli (1191–1218?)*, Università di Torino, Memorie e studi geografici, 10 (Turin: Fanton, 1976), especially pp. 5–18, 117–25.

93 Hartmut Kugler (ed.), *Die Ebstorfer Weltkarte*, 2 vols (Berlin: Akademie Verlag, 2007); a much shorter but well-illustrated account of the map is Birgit Hahn-Woernle, *Die Ebstorfer Weltkarte* (Ebstorf: Kloster Ebstorf, [1988]). Studies of particular points of detail are in Hartmut Kugler and Eckhard Michael (eds), *Ein Weltbild vor Columbus* (Weinheim: VCH, 1991), among them the important work on its date by Armin Wolf (pp.

54–119), Renate Kroos (pp. 223–44) and Horst Appuhn (pp. 245–59). Of Wolf's work there is an abridged version in English, 'News on the Ebstorf World Map: Date, Origins, Authorship', in Monique Pelletier (ed.), *Géographie du monde au Moyen Âge et à la Renaissance,* Mémoires de la Section de géographie, 15 (Paris: Ministère de l'éducation nationale, Comité des travaux historiques et scientifiques, 1989), pp. 51–68. Barber, 'Medieval Maps', pp. 23–7.

94 G. Haslam, 'The Duchy of Cornwall Map Fragment', in Pelletier (ed.), *Géographie du monde,* pp. 33–44; Barber, 'Medieval Maps', pp. 19–23.

95 Discussed by P. D. A. Harvey, 'The Holy Land on Medieval World Maps', in Harvey 2006, pp. 243–51.

96 Thomas O'Loughlin, 'An Early Thirteenth-Century Map in Dublin: A Window into the World of Giraldus Cambrensis', *Imago Mundi,* 51 (1999), pp. 24–39; Barber, 'Medieval Maps', pp. 14–16. It is interesting, but probably not significant, that the map of Europe is in an early-thirteenth-century manuscript that was given in 1438 to the Hereford vicars-choral; it is now Dublin, National Library of Ireland, MS 700.

97 See n. 86, above.

98 P. D. A. Harvey, 'Matthew Paris's Maps of Britain', in P. R. Coss and S. D. Lloyd (eds), *Thirteenth Century England: IV* (Woodbridge: Boydell, 1992), pp. 114–16; P. D. A. Harvey, 'Matthew Paris's Maps of Palestine', in Michael Prestwich, Richard Britnell and Robin Frame (eds), *Thirteenth Century England: VIII* (Woodbridge: Boydell, 2001), pp. 173–5.

99 Barber, 'Medieval Maps', pp. 32–5.

100 P. Barber, 'The Evesham World Map: A Late Medieval English View of God and the World', *Imago Mundi,* 47 (1995), pp. 13–33.

101 P. Barber and M. P. Brown, 'The Aslake World Map', *Imago Mundi,* 44 (1992), pp. 24–44; Barber, 'Medieval Maps', pp. 35–8.

102 Westrem 2001 discusses each of the map's principal written sources in general (pp. xxvii–xxxiii) as well as referring individual inscriptions to a particular source, and may be consulted for many of the points mentioned in this chapter.

103 e.g. Harvey 1991, p. 90.

104 Both Miller 1896, p. 33, and his facsimile of the map misread this inscription in trying to make sense of it; it is given correctly by Bevan and Phillott 1873, p. 74.

105 Bevan and Phillott 1873, p. 77.

NOTES

106 Miller 1896, p. 32.

107 Bevan and Phillott 1873, p. 15; Miller 1896, p. 7.

108 Miller 1896, p. 48.

109 Westrem 2001, pp. xxix–xxx; Miller 1896, p. 49; Crone 1954, pp. 18–22; Crone 1965, pp. 450–2, 455–6.

110 Miller 1896, pp. 52–4; Crone 1954, pp. 14–15, 22–3.

111 Miller 1896, p. 17.

112 Kline 2001, pp. 165–90; Naomi Reed Kline, 'Alexander Interpreted on the Hereford Mappamundi', in Harvey 2006, pp. 167–83.

113 This is surely the identity of 'Cadan', neither a seriously misplaced Caen (as Bevan and Phillott 1873, p. 170; Miller 1896, p. 18) nor a seriously miscopied Salisbury (as Crone 1954, p. 9). Though its identification with Camelot is first recorded in the mid-sixteenth century (Geoffrey Ashe, *The Quest for Arthur's Britain* (London: Pall Mall, 1968), p. 156), it was probably a long-standing tradition.

114 Denholm-Young 1957, pp. 312–13. That Kirkham was one of the many places that Edward I visited (Crone 1954, p. 9) seems unlikely to be relevant.

115 *Transactions of the Woolhope Naturalists' Field Club* (1868), p. 238; (1869), p. 152.

116 Miller 1896, pp. 4–5.

117 Miller 1896, p. 5 n. Among further alterations that Miller made silently on 3 and in his text are 'uiuun*t*' for 'inueni' in the inscription on the beasts like minotaurs, and the substitution of *th* for *þ* in 'Muneþ' and 'baþe'.

118 Bevan and Phillott 1873, preface.

Illustrations

Illustrations from The Folio Society's reproduction of the Hereford map (2010) are not listed.

ILLUSTRATIONS

ILLUSTRATIONS